T5-ARJ-323

THE PEDESTRIAN AND CITY TRAFFIC

to Johanna and Graham

The Pedestrian and City Traffic

Carmen Hass-Klau

Belhaven Press
(a division of Pinter Publishers)
London and New York

388.41
H35p

© Carmen Hass-Klau, 1990

First published in Great Britain in 1990 by
Belhaven Press (a division of Pinter Publishers),
25 Floral Street, London WC2E 9DS and
P.O. Box 197, Irvington, New York, USA.

All rights reserved. No part of this publication may be
reproduced, stored in a retrieval system, or transmitted by any
other means without the prior permission of the copyright holder.
Please direct all enquiries to the publishers.

British Library Cataloguing in Publication Data

A CIP catalogue record for this book is available from the
British Library

ISBN 1 85293 121 3

Library of Congress Cataloging-in-Publication Data
Hass-Klau, Carmen.
    The pedestrian and city traffic / Carmen Hass-Klau.
        p.   cm.
    Includes bibliographical references (p.   ).
    ISBN 1 – 85293 – 121 – 3
    1. Pedestrians. 2. City traffic. 3. Transportation, Automotive.
4. Streets – Planning. 5. City planning. I. Title.
HE336.P43H37 1990
388.4′1 – dc20

Filmset by Mayhew Typesetting, Bristol, England
Printed and bound in Great Britain by Biddles Ltd.

Contents

UNIVERSITY LIBRARIES
CARNEGIE MELLON UNIVERSITY
PITTSBURGH, PA   15213-3890

List of figures

List of tables

Preface

The completion of this book has only been possible because of the help and support of numerous professionals in both countries, Germany and Britain. But my sincerest thanks have to be directed to my family, Graham and Johanna, who had to cope with all my ups and downs of simultaneously being a mother, consultant and writing a book.

I would very much like to thank Professor Peter Hall, who assured me that there was no need to panic or to despair and that I only had to do what he had suggested. In the end I did not quite do this and I hope he will forgive me. I had many helpful and inspiring discussions with him. I also have to express my warmest thanks to Sir Colin Buchanan who was kind enough to spend many hours discussing the interesting period of the 1960s in Britain, and I am very pleased that he was willing to write a Foreword.

I would further like to thank Professor Peter Koller for the intensive and very helpful correspondence with me. My thanks have also to be extended to Senior Lecturer Boudewijn Bach at the University of Delft for his comments on the Dutch section in Chapter 10.

I want in particular to thank Dr Heiner Monheim at the Ministerium für Stadtentwicklung, Wohnen und Verkehr des Landes Nordrhein-Westfalen and Dr Hartmut Keller at the Bundesanstalt für Straßenwesen for reading Chapter 10 and making very useful comments, Dr Raabe at the Akademie für Städtebau und Landesplanung in Berlin for sending me a crucial article, Peter Heidebach of the Stadtplanungsamt München, Regierungsrat a.D. Kern of Wolfsburg, Bärbel Pook of the Stadtplanungsamt Hamburg and Frau Kier at the Amt für Denkmalschutz in Cologne. Helmut Hass helped as always to find some obscure German references in Berlin. I am also grateful to the main librarian of the Bundesforschungsanstalt für Landeskunde und Raumordnung whose help in finding some of the historic articles was extremely valuable.

In Britain I am indebted to Mr Lancaster at the Garden City Museum Letchworth, John Roberts, from TEST, for reading Chapter 11, the Department of Town and Country Planning at the University of Manchester for giving me access to the Unwin collection, and many other librarians at various British libraries, the Rowntree Trust and the Hampstead Garden Suburb Society. I have also to thank many officers in several planning departments for sending me relevant material. Reading Dr Miller's thesis was very helpful and without his research several gaps could not have been filled.

I am extremely grateful to Inge Nold who was in charge of the draw-ings and gave advice on the quality of the photographs. I would also like to thank the team of Environmental and Transport Planning for their patience.

I cannot include all the many others who helped in one way or another to finish this book. But despite all the help I am solely responsible for all the errors and omissions.

Carmen Hass-Klau
Brighton

Foreword

Ever since an alternative method of land transport to Shank's pony was first invented the not-so-fortunate travellers on foot have been at risk. Riding on a horse was probably the first innovation, but who nowadays does not experience a pang of fear when obliged to walk past a mounted horse at close quarters? Roman chariots drawn by thundering horses must have terrified the lightly armed opponents on foot, though perhaps not quite as much as Queen Boadicea did in her chariot with long knives attached to the wheel hubs. City streets crammed with horse-drawn carriages and carts, slippery with every kind of muck, stinking to high-heaven as well, must have brought death and injury to many pedestrians trying to cross. Nor were railway trains free from blame, as in the case of the great statesman William Huskisson who, in the year 1830, not only attended the opening of the Liverpool and Manchester Railway but was killed when walking across the track.

Trouble on a much more serious scale came with the arrival on the scene of the motor vehicle. It quickly established its supremacy in door-to-door transport for people and goods and then began its penetration of pedestrian-crowded urban streets. The results in almost every city in the world have been a mixture of benefits and disaster. Benefits in the sense that more is done, more is made, and life goes faster; disaster because so many people are killed and injured, noise and pollution and other side effects are rife and because the influence of the motor vehicles calls into question the very principles of city design and the organization of transport. The fundamental question concerns the extent to which cities should be changed, even rebuilt, to accommodate the popular demand for freedom to use motor vehicles (especially the car) for every conceivable transport need. The argument has been going on for nigh on a century, but there now seems to be a feeling in the air that as far as rebuilding the cities is concerned 'enough is enough'.

In this book Dr Hass-Klau performs a public service first with an historical review of the development of ideas (especially in street design) for the safeguarding of pedestrians, and secondly with invaluable suggestions as to how we should proceed. Although reference is made to practice in France and the United States, the most deeply interesting reference lies in the comparison between German and British approaches. Here was a curious interchange of ideas spread over many years and not significantly deflected by events in Germany under the Third Reich.

However there is no doubt that the message of the book which will most concern us in Britain concerns the plight of urban pedestrians and what can be done to relieve it. In the book Dr Hass-Klau pays generous tribute to the official British report 'Traffic in Towns' published by Her Majesty's Stationary Office in 1963. It is certainly the case that 'Traffic in Towns' identified all the anti-environmental effects of motor traffic, from death on the road to pollution, and presented them as a major social problem on a world-wide scale. Certainly it was demonstrated how some of these adverse effects could be ameliorated by such measures as removing traffic from shopping streets and the technique referred to as 'environmental management'. Certainly warning was given of the 'heroic act of self-discipline' that would be needed (especially by drivers) as the number of vehicles in circulation continued to increase. But it would be erroneous to attribute these thoughts as exclusive to 'Traffic in Towns'.

I myself, as leader of the team which produced 'Traffic in Towns', and while we were writing it, travelled extensively in West Germany to see what ideas were developing there and what I could learn. To give but one example, I saw that pedestrianization of shopping streets was not just an idea in the sky but was being vigorously applied, as it had been in the intervening years, with the result that coherent networks of pedestrian streets now exist in German cities on a scale unmatched anywhere else in the world. I might add that I also travelled widely in the United States and saw there the beginning of the move towards unhurried steady-speed disciplined driving, which is now characteristic of traffic on most North American highways in contrast to the speed-obsessed drivers who career along European roads.

The contribution which Hass-Klau now offers in this book marks a further important step in the control of traffic in towns. To put it simply she is pointing out that while there is a limit to the amount of motor traffic any town can accommodate if acceptable environmental conditions are to be maintained, there is nevertheless a not considerable amount of traffic that must be admitted if the life of the town is to continue. This does not prevent some urban streets (e.g. shopping streets) from being closed to motor traffic altogether, but there are many other streets where pedestrians and motor vehicles have to live together. These are the streets, and there are many of them in British towns, where almost intolerable conditions exist for pedestrians. To deal with this problem Hass-Klau advocates the technique now increasingly in use in German towns and village under the name of *Verkehrsberuhigung*. This translates, perhaps a little uneasily, into English as 'traffic calming'. But the objective is perfectly clear – to allow pedestrians and a quantum of vehicles to mix safely in the streets, and this is achieved by means of comparatively inexpensive street designs which positively curb the actions of too-fast and uncaring drivers. This fills a glaring gap in the present system of managing urban traffic; there is wide scope for its application in British towns. One cannot but also hope that a technique on the same lines will soon be applied to traffic circulation on the whole of our road system with the simple objective of 'calming it down'.

Colin Buchanan
Oxford, October 1989

1 The differences: do they matter and are they explainable?

This book gives a detailed picture of how planners, politicians, residents and transport engineers in three societies, Britain, Germany and the United States, reacted to one of the most powerful inventions of the late nineteenth century, the motor car. Misjudgements of the potential growth of motor vehicle ownership and its adverse effects had serious repercussions in the coming decades, primarily in the dense urban areas.

Underestimating the importance of public transport as a real alternative to the motor car in urban areas, first by the United States and several decades later by Britain, has been a disturbing feature. Of the three countries, only Germany seems to have struck a better balance.

Not surprisingly, at the beginning of the twentieth century, conflicts already occurred between the weaker road participants (pedestrians and cyclists), the existing urban fabric and the motor vehicle. A more comprehensive comparison between Britain and Germany shows that both countries developed specific patterns and had different attitudes towards road transport. Far more has been invested and planned in Germany, whereas Britain has shown not so much a lack of foresight in planning but in investment in road transport. This major difference has had very visible effects on today's urban structure and transport situation.

The demand for restraint of motor traffic had different motives in the two countries, and is not such a new idea as is often assumed. While in Germany even in the 1920s and '30s the protection of historic inheritance was a decisive motive, in Britain that was not the case. Questions of traffic restraint were however raised in connection with road safety and later in the 1960s as a means of improving the urban environment.

The turning point of nearly unlimited promotion of car use in urban areas took place in Germany during the 1960s and '70s, whereas the Buchanan Report had already warned in the early 1960s against the adverse effects cars could have in urban areas if they were not controlled. Although even in Britain the report was misunderstood and largely not put into practice, the wave of protest against road-building occurred earlier there than in Germany. As a whole, Britain has shown a brilliance of ideas in restraining motor vehicles which was lacking in the Federal Republic. At the beginning of the 1970s, serious discussions in Germany on how to restrain and control the motor vehicle were slowly transformed into reality. Britain seems to have followed this example, but with a considerable time-lag.

Working as a transport and planning consultant in Britain, I realized that there were substantial differences between urban road transport in the two countries. Why is it that German pedestrianization appears to have such a completely different quality which reinforces the historic urban functions, in contrast to the British equivalent? Why was it that Germany had invested heavily in new urban public transport systems while Britain deregulated its bus industry and had neglected investing in urban public transport for decades? And why was it that German transport planners implemented restrictive measures against motor vehicles in residential areas and very recently on main roads, while only in the last couple of years have British local and central government appeared to be interested in this approach?

Obviously one could explain these differences by the past and current political and socio-economic systems in the two countries, although one has to be aware that Conservative governments have ruled in Germany too. The division of the federal political structure of West Germany into States (or Länder), which have the planning and most of the transport powers, will also provide some valuable explanations. There are substantial variations between the two countries in the financing of pedestrianization or any other physical road changes. Yet, a more comprehensive understanding may be buried deep within the countries' specific history of urban road transport planning, the objectives for designing street layouts, the importance of preserving the urban environment, the consequences of road safety, the influence of the road lobby, particular political forces or individual decisions. Only an in-depth historical study of both countries would enable me to find patterns, ideologies and practices. These could have developed differently very early on in each country because of specific circumstances, and these differences may have continued and could possibly explain the situation today in Britain and Germany.

This book is a comparison primarily between Britain and Germany and it had to be a comparison because in all my professional activities I have been confronted with the whole range of variations between these two countries. A third country, the United States, has been added mainly in order to help understand in particular those historical issues which influenced Britain and/or Germany.

Many of my colleagues see transport planning largely as an expression of present political decisions which have to be made according to current demands and interests. There is often a total lack of any historical understanding of past transport and planning ideologies and decisions and how these were interrelated. Reading the relevant transport journals one becomes convinced that road transport policy is ruled by engineering aspects only. But is that really the case? Are both transport planning and their policies as ahistoric as they appear to be on the surface? Relatively few experts are involved in carrying out research into planning and transport history and appreciate its importance in understanding current planning and transport concepts. This is even more valid for transport planning. Therefore the question arose whether the modern German approach of controlling motor traffic, which seems so unique today, was simply the logical outcome of a different understanding of cities, urban life in general, past ideas, concepts and experiences of transport and planning issues. The processes I analysed are a complex combination of socio-economic, political, cultural and individual decisions. I was also interested in whether and to what degree both countries influenced each other on this issue, or whether each society coped in its own specific way with motorization.

The main question was whether the two societies developed policies to protect pedestrians and residents from motor traffic and, if they did, what kind of policies were they, and had they already applied them when wheeled traffic started to become a nuisance in urban areas? I was mainly interested in the period when road construction and road planning began to be more seriously developed to cater for the needs of motorization. How did people living along such roads react to this new form of transport? In what ways did architects and planners adjust housing developments to the motor vehicle? Did street layouts change in residential areas and in town centres? What were the new ideas on how to improve or cope with traffic congestion? Are misjudgements of the past with reference to motor use still paid for today? Most important of all, what were the actions and reactions in Britain in comparison with Germany? Most of these questions seemed unanswered. My final conclusions may not satisfy everybody and there may be many who disagree but I still feel that it is worth discussing these issues in order to be able to make changes.

When starting writing this book the most important topic for me was a relatively recent German transport policy, known as 'traffic calming'. I wanted to understand why it was such a big issue in Germany and not in Britain. In order to clarify what I mean by it, the next section will give some basic information and a definition of traffic calming but in Chapters 11 and 12 we will come back to this issue.

Traffic calming

Traffic calming is by no means a German invention. It was originally developed in the Netherlands under the well-known term *Woonerven* during the late 1960s and early '70s. Since then it has been developed much further and successfully applied in many other European countries. I have been studying primarily the German approach because of my background and knowledge.

I have always seen German traffic calming as a continuation of pedestrianization with which I was very impressed when first seeing it after years abroad. What impressed me most was how the Germans were able to reclaim urban space for pedestrians so successfully. This has also been the case with traffic calming, which was at the beginning primarily applied in residential areas but is now taking over whole cities. As with pedestrianization, road space is given back to the cyclist and pedestrian. It is an attempt to mix the different transport modes and create a form of peaceful coexistence between them which according to the character of the road will vary. The result is that in most cases the urban environment is considerably improved.

Traffic calming is both a planning and a transport policy and may in the future become a new way of life in built-up areas. If we are serious about the cultural inheritance of our towns and villages it may well be the only way forward to cope with the increasing number of motor vehicles which have been forecast.

In the English language, 'traffic calming' is becoming a more familiar term, and Oxford University Press is considering including it in the next edition of the Oxford English Dictionary after it was published in *The Times*. In professional circles it is used quite frequently and even the British Department of Transport is, after some reluctance, referring to it. However few people know its meaning and when writing this book there was no precise definition universally accepted.

Traffic calming can be used both in a wider or in a more restrictive sense. In a *wider* sense it may be defined as an *overall transport policy* concept, which includes, apart from a reduction of the average motor vehicle speed in built-up areas, a strong promotion of the pedestrian, public and bicycle transport. It also involves different restrictive measures against motor vehicles according to defined needs of the built-up environment. Road pricing or parking restrictions may be appropriate in some urban areas but not in others. Answers have to be found for the heavy lorry transport which will become heavier and will further increase. A rethink of car taxation and company car allowance should also be a crucial part of such a policy. Traffic calming in a wider sense will include a combination of transport and, most important, land-use policies which will reduce the number of motor vehicle trips, and even more desirable the *need* for motor vehicle trips in built-up areas. Traffic calming is a bundle of transport policies intended to alleviate the adverse environmental, safety, and severance effects motor vehicles have been and will be creating.

However traffic calming is far from being a witch-hunt policy against the car. It will be the struggle for the 'emancipation' of the pedestrian, the reclamation of public and cycle transport and the preservation of the historic environment.

In most of this book, traffic calming has been defined in a much narrower sense which may well be the first step towards an overall traffic calming concept. It is seen as a policy to reduce motor vehicle speeds from a maximum of 50kph to a maximum 30mph in urban areas. In some residential streets a further reduction to even lower speeds may be desirable. This type of traffic calming has three main objectives:

1 to reduce the severity and the number of accidents in built-up areas;
2 to reduce air and noise pollution;
3 at the same time to 'improve' the urban street environment for non-motor users.

The *road-safety* objective aims to reduce the speed of motor traffic to 30kph or below in built-up areas. Motor vehicles generally drive too fast on urban roads. The maximum speed is 50kph or 30mph but in reality many drive faster than that. We know that if a car has an accident with a pedestrian at 70kph, the likelihood that the pedestrian will be fatally injured is 83 per cent. If the car drives at 50kph the likelihood will still be 37 per cent, while at 25kph it will be 3.5 per cent. Thus the more we reduce the maximum speed the more will the pedestrians have a chance to survive an accident.

A reduction of motor speed is normally carried out either by physical changes of the carriageway, such as chicanes, parking at right-angles, bottlenecking, raised crossing, and/or traffic signs which restrict the speed of motor traffic.

The *planning* objective is difficult to define in quantitative terms, as improvement of the urban environment will be judged differently by individuals. Even so, there are some criteria which appear to improve the urban environment for non-motorized street users, such as a well-designed bicycle path, adequate and uncracked pavements, trees, flower-beds, benches, or playground facilities.

The *environmental* objective implies at best a reduction of the levels of air pollution and noise level which are inflicted by motor vehicles. Both have an impact on the improvement of the urban environment.

Yet traffic calming is wrongly understood if it is purely applied as a road traffic engineering policy. There is a danger if only a few residential streets are redesigned into *Woonerven* or are cluttered with speed humps, flower beds or other constructions of the vertical or horizontal alignments. This style of traffic calming will reduce the number of motor vehicles or the average motor vehicle speed from the streets which have been treated, but traffic will redistribute itself into other residential streets. We can only be serious with traffic calming if we are willing to 'hurt' all motor vehicle users and give substantially more ground to the weaker road participants.

The structure of the book

The structure is laid out chronologically. The book begins with the earliest ideas on protective road transport policies in several countries. Some chapters, like 3 and 4, deal with Germany and Britain together because of the extremely close links between them. The following chapters alternate between Britain, Germany and the United States, each covering roughly the same time periods.

Chapter 2 starts with a review of the first attempts to protect pedestrians from wheeled traffic. It includes considerations of important aspects of traffic-controlling policies, such as the development of traffic signs and traffic management. It also gives a short overview of futuristic traffic ideas, suggested around the same time, which dealt with traffic separation. The chapter ends with a short section concerning the degree of motorization in Europe and the United States during the 1920s and '30s.

The German and British attempts to cope with the invasion of the motor vehicle are discussed in the following chapter. It starts with the discussions about the different schools of thought on German street planning at the end of the nineteenth century and gives the equivalent development in Britain. This chapter shows already the differences in road standards and the importance of roads in the two countries. It includes a section on the arrival of the motor vehicle and how the two societies reacted to it. The last part describes the road-safety discussions and the policies during the 1920s and '30s.

Innovative residential street layouts developed during the British and German Garden City movements are the topic of Chapter 4. It analyses the deliberate planning attempts to protect pedestrians from the adverse effects of wheeled traffic. It considers too the relevant transport ideas which had been valid for the city centres.

In Chapter 5 I argue that the Weimar Republic was not producing any new ideas in road planning and that some of the ideological variations between the radical and traditionalist architects and planners were not so different in terms of road design as often is made out. Some of the fictional ideas of the more radical architects would have had quite dramatic effects on the traditional urban structure as we in part could experience after World War II.

The degree of motorization in the United States and the first attempts to develop new street layouts to cope with motorization are discussed in Chapter 6. The most interesting aspect is the European influence on the street layout developed in the new housing estates in Radburn and the Greenbelt cities. The well-known design of separating different traffic modes which was perfected during the 1920s and the 1930s in the United States was introduced back into Europe in later decades.

Chapter 7 gives an insight into the ideology and realities in terms of design, planning and implementation of roads during the Third Reich. Some attention is given to the construction of motorways and to the development of the Volkswagen. But the main part of the chapter includes specific examples of residential planning which were designed to protect residents from the adverse effects of motor vehicles. In addition, the different approaches on how to cope in the city centres with the new demands of motor traffic are set out.

The discussions on road safety and environmental questions by Alker Tripp and Colin Buchanan are referred to in Chapter 8. It gives a detailed account of the background to the Buchanan Report and its impact in the following decade.

Chapter 9 explains the West German path into mass motorization and sets out the first policy to cope with it. It also points out that a few of the car-restricting ideas of the Third Reich were still being discussed during the 1950s but had little effect on the overwhelming road transport policies in favour of car use. However, it becomes clear that, despite vigorous attempts, the vision of total motorization was not a real option in Germany.

Chapter 10 continues with Germany mainly because it helps to understand better the very recent policies on pedestrianization and traffic calming in Britain, which have been in part influenced by Germany and other Central European countries. This chapter is devoted to the history of pedestrianization and traffic calming, its ideological background, successes and failures in Germany since the 1970s.

The transport policies in Britain with reference to protecting pedestrians and residents from the adverse effects of motor traffic during the 1970s up to 1989 are considered in Chapter 11. It shows the various attempts in Britain to implement pedestrianization and traffic-calming policies in residential areas. Their successes and failures have been related to the political and economic situation of Britain but also to the acceptance by the affected residents. Included in it are the current discussions and the first practical attempts at traffic calming in Britain.

Finally, the last chapter summarizes the main conclusions of the book which tries to thread together and interpret the main variations between Britain and Germany.

2 A review of the history of pedestrianization and other protective ideas and policies before World War II

Background

This chapter is a review of the major historic transport developments which include the first attempts at separating traffic from pedestrians. It will touch on some basic transport and planning ideas which were being discussed by the end of the nineteenth and at the beginning of the twentieth century. However more detailed material on these issues is presented in the following chapters. As outlined in Chapter 1, this book is concerned with the historic development and interconnecting links between Britain, Germany, and to a lesser extent the United States. In this chapter, additional countries have been included on most issues so as to emphasize particular ideas.

Roads and streets developed mainly according to the needs of overland travel. The existing geography and the seasons played a major role in the form and route that roads would take. It would probably be safe to assume that a large majority of roads had been used for centuries. In most European countries serious road-building had taken place during Roman times. During the following centuries most Roman roads had been neglected and often merely tracks or bridle paths remained. In medieval times, few transport modes were available. Overland travel was either carried out on horseback or on foot. Goods were transported by pack horses. Although various forms of carriage had been developed mainly from the late sixteenth century onwards, they only became common during the last quarter of the eighteenth century (Webb and Webb 1913: 72). With the increase in wheeled transport, tracks and bridle-paths had to be changed into roads which ideally would work best with smoother and harder surfaces.

In built-up areas, walking was the usual means of getting around, to work, or to attend social activities. The rich could afford to be carried

Figure 2.1 Venice: typical 'street'.

in sedan chairs, or to use carriages and horses, and the latter two were mainly used to overcome long distances. Only a few instances of conflicts between pedestrians and wheeled traffic are known, for instance the congestion of the streets in Rome which led Julius Caesar to ban carts and chariots from the city between sunrise and sunset (Breines and Dean 1974: 14). The Forum of Pompeii could only be used by pedestrians and the streets leading to the Forum ended as cul-de-sac roads.

The most splendid example of early separation between traffic and pedestrians can be found in the city of Venice which developed from the fifth century onwards (Mumford 1961: 369). The dredging of canals was

Figure 2.2 Venice: footpath access to houses.

used to fill in the land, and gondolas (first mentioned in 1094) became the most important transport mode. Even today, Venice is the most perfect example of a city in which transport modes are separated.

Some of the Dutch cities were planned and built in a similar way to Venice. There, the building of canals started in the sixteenth and early seventeenth century. But, in contrast to the Venetian canals, the Dutch canals had sidewalks which allowed even carriages to enter the city. There are some footpaths next to some of the canals in Venice but they are more the exception than the rule (see Figs 2.1 and 2.2). Yet the basic principle of traffic separation in the Dutch cities was similar to Venice. The waterways were used to carry the main flow of freight goods.

The development of pavements and arcades

Pavements protected pedestrians from wheeled traffic and were used as long ago as Roman times, but had been forgotten until the late seventeenth century. After the Great Fire of London in 1666, pavements were provided in all the newly constructed streets. In France and Germany pavements were not known until the middle of the eighteenth century, and then only in the 'better' streets. Haussmann and Alphand included pavements in all streets in Paris (Geist 1983: 62–4).

During the eighteenth century, *covered passageways* which later became famous for their glass-covered roofs were firstly built in France, spreading later to all large European and some overseas cities. Arcades were the first planned attempt in congested urban areas to separate pedestrians from wheeled traffic. They were constructed mainly in and for the city centres, and could be described as linear or crosswise shopping centres to protect pedestrians from weather and traffic. The question remains open whether arcades were the predecessor of pedestrianization or of shopping centres; surely they were both. The main economic function of arcades was to follow a different style of shopping which had developed from the middle of the eighteenth century.

Changes in the social behaviour of the public – in particular the growth of shopping as a form of communication for the upper-middle classes – may have been an important motive for their construction. The protection from traffic and weather was secondary though an important aspect. Geist believed that arcades came into fashion because streets were dirty and dangerous to walk in.

The amount of traffic on the narrow Parisian streets took on dangerous and threatening proportions at this time (Geist 1983: 62).

The most famous arcades were the Burlington Arcade in London, built 1819, covering about 2,000sq.m and being 180m long; the three-storey-high Kaisergalerie in Berlin was built much later, just after the foundation of the new Empire, in 1871–3, covering 4,750sq.m. The Friedrichstraße Arcade was nearly twice as large and built as late as 1904. Without doubt the most famous arcade was the Galleria Vittorio Emanuele in Milan which opened in 1867 (4,200sq.m), influencing the style of many other arcades all over the world. Most of them were between 8m and 14m wide and had a substantial length; the one in Milan in 195m long.

Many of the great arcades were built between 1820 and 1880. Geist pointed out that before World War I, with new building regulations, arcades in their nineteenth-century form disappeared.

However arcades were not the only attempts to separate traffic during the nineteenth century.

Traffic separation: a necessary measure to protect the character of parks

The Radburn principle, the twentieth-century concept of separating pedestrians from motor vehicles, was derived, at least in the Anglo-Saxon countries, from the road design of landscape architects. Public parks were designed around the middle of the nineteenth century. At about that time there was a strong movement to create open spaces as a necessity for the well-being of the public. Thus in *Britain*, royal parks were opened to the public or new parks designed.

The open-space movement was also connected with the struggle to obtain the rights of public footpaths. The 1815 Highways Act allowed the closing down of public footpaths, an endeavour which caused public indignation (Webb and Webb 1913: 203). The demand for maintaining public walks, e.g. the walk along the Thames embankment, was an attempt to modify the 1815 Act. In the new industrial-growth areas public rights of way were arbitrarily abolished. As a result in many cities, e.g. in Manchester, societies were formed to protect public footpaths. In Liverpool the St James Walk, formed by the City Corporation of Liverpool, was partly enlarged into a garden but had been neglected already by 1833 (Olmsted *et al.* 1928).

The *Germans* also had a vigorous open-space movement which continued during the period of National Socialism. A number of parks were built and opened for the public somewhat earlier than in Britain. The first municipal park in Germany was Friedrich Wilhelmsgarten in Magdeburg designed in 1824. Friedrichshain in Berlin was laid out for the public in 1840 and Birkenhead Park, the only park of any significant size built outside London, was finished in 1845. Joseph Paxton, who had designed Birkenhead Park, separated its roads according to the different modes. The park included a separate road for carriages and two completely separated footpaths.

The first public park in the *United States* was Boston Common which was established as early as 1634. It took until 1875 before the first park commission for Boston was created, about twenty years later than in New York (Reps 1965: 333). Fredrick Law Olmsted and Calvert Vaux who designed the Central Park in New York after 1857 (Olmsted *et al.* 1922: 98) used a similar system to Paxton, though somewhat more sophisticated. A completely independent network for each mode with overpasses at crossings was designed, also including both a major through road and a wide parkway which were supposed to become part of the metropolitan system of parkways and special highways (Olmsted *et al.* 1928; see Fig. 2.3). Olmsted had visited Birkenhead Park in 1850 and had been rather impressed by the layout. He specifically mentioned the road and path layout of the park in his diary (Olmsted *et al.* 1922: 125). There is a strong possibility that he copied the road layout of Birkenhead Park when designing Central Park. Olmsted in his later work designing new housing areas and boulevards very often made use of the idea of separating traffic, e.g. for the boulevards in Hyde Park, Chicago or to some extent in Riverside, Illinois (Scully 1969: 88; see Fig. 2.4). Olmsted influenced a relatively new subject, landscape architecture. Landscape architects became rather important in designing street layouts for housing areas. Hegemann wrote that in the United States there were several housing estates which were built similar to British and German garden suburbs. The division of main roads and footpaths and the grouping of housing was part of their design, e.g. Black Rock in

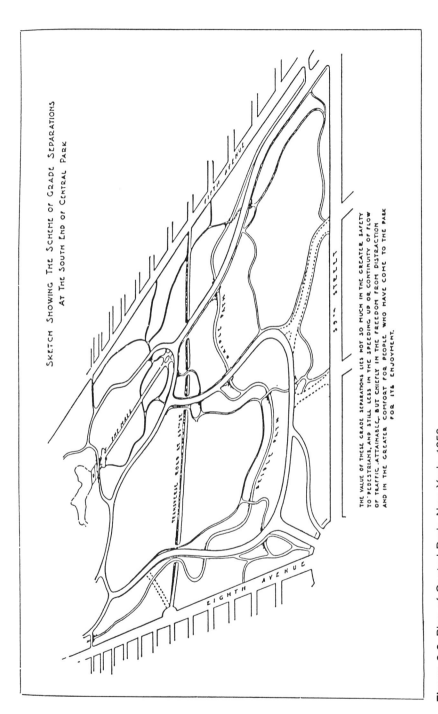

Figure 2.3 Plan of Central Park, New York, 1858.

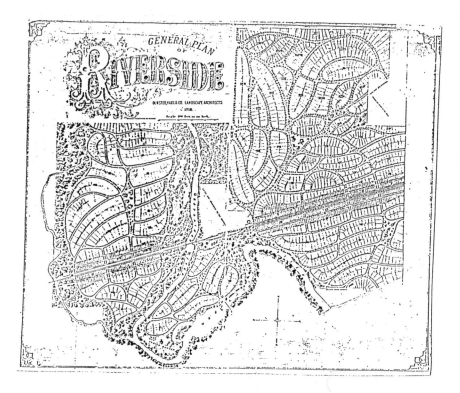

Figure 2.4 Plan of Riverside, Illinois, 1869.

Bridgeport and Roland Park, Baltimore designed by Olmsted (Hegemann 1925: 119–24).

According to Mumford, Clarence Stein was influenced by the road design of Central Park when he developed plans for the housing estate in Radburn. However it was much simpler than Mumford pointed out: Clarence Stein and Henry Wright, who was a landscape architect, implicitly followed the tradition of street design of the American landscape architects, and made it more distinctive in order to cope with motor traffic. I have argued in Chapter 4 that the separation of roads and footpaths first partially developed in Port Sunlight may have been influenced by Paxton's design of Birkenhead Park.

New forms of street layout to protect people from wheeled traffic, which were used firstly in parks and later in newly designed housing estates, were not the only way of restricting traffic. In the large metropolis, traffic congestion had become a problem with the increase of wheeled vehicles.

Traffic regulations to ease congestion in the large cities

From the middle of the nineteenth century, traffic congestion in many large cities had become a serious problem. But long before many main streets in the large cities became congested. Leonardo da Vinci was the first known 'planner' to suggest the separation of pedestrians from heavy traffic arteries to solve the traffic problems of Milan during the fifteenth century (Mumford 1961). His idea was to put the traffic underground.

The City of London and parts of Westminster were already known to be notorious for their traffic congestion. Kellett wrote that

London's traffic congestion in the 1860s, however, had reached crisis proportions (Kellett 1969: 317).

Traffic congestion was not only confined to London. Similar statements were known from other major cities.

The unpleasantness of traffic congestion was partly caused by the large demand for horse buses and horse trams. By the end of the nineteenth century, horse trams started to be replaced by electric trams, which improved the street conditions slightly (see Fig. 2.5). Yet the speed of replacement was different in each country. In Germany, electric trams were introduced very rapidly. It occurred rather slowly in Britain because of special legal requirements which made new investments something of a gamble for the entrepreneur.

Traffic congestion was further increased by two major inventions: firstly, the *bicycle* developed during the second decade of the nineteenth century, which was in common use from 1895 onwards, and secondly, the *car* developed in 1885 by Carl Benz and Gottlieb Daimler. Bicycles were used mainly by the young and lower income groups of society, and they became very popular in Britain and in many other European countries. The use was highest in the Netherlands, Denmark, Belgium and Germany where it peaked during the 1930s. In Germany about 30 per cent of all street users were cyclists in the early 1930s. Although the bicycle could hardly have been a major cause for traffic congestion, the trams, motor buses and freight traffic certainly were.

Road congestion could be eased by traffic regulations. They were invented in their modern form by the American William Eno who published in 1901 a set of General Highway Traffic Regulations (Eno 1929: 3). They were soon accepted and introduced in New York in 1903 and revised several times in the following years. Other US cities adopted them and Paris introduced the same highway code in 1912.

In the United States, at about the same time, the official Council of National Defense Code of General Highway Traffic also included in its definition a description for 'restricted zones' which sounds astonishingly modern. A restricted zone was that part of a roadway on which pedestrians were allowed but from which all vehicles except tram cars

Figure 2.5 Street scene in Liverpool, 1908.

were excluded. The existence of this code implies that it must have been used.

One of the most important features of the highway code was the one-way street regulation, first introduced in New York in 1907, Boston 1908, Paris 1909 and Buenos Aires 1910.

Other elements were manual and technical traffic control of traffic flows, later replaced by traffic lights. Eno believed that the technical device of controlling traffic originated in England and was later adopted in France and used in New York for the first time in 1902. Indeed as early as 1868 there was a street crossing signal in London which was a stop and caution sign, similar to the signs used at railway lines. It also included a red and green light during the night and was installed outside the Houses of Parliament. The effect of the red and green light was achieved with gas which unfortunately blew up after working for only two hours.

According to Tripp the first three coloured *traffic lights* were set up in New York in 1918 (Tripp 1950: 258). In Britain such lights were first used for railway operations, for instance the Liverpool Overhead Railway Company (1920) and the London and North Eastern Railways (1923). Shortly after 1923, they were also used for roads. Paris had its first traffic lights in 1922 and traffic lights were also installed in Berlin (Potsdamer Platz) in 1924.

The rotary system – *roundabout* – was implemented first around Columbus Circle in New York in 1903, and adopted for the Place de l'Etoile in Paris 1907. Five years later it was seen for the Rond Point on the Champs Elysées. Eugène Hénard had drawn his own roundabout – 'merry go round' – in 1906 which included pedestrian subways as crossings (Le Corbusier 1924: 121). In 1926 Piccadilly Circus was regulated as a roundabout. One year later roundabouts were used in seven other places in London (Eno 1929).

According to Eno, before 1903 there was no attempt made to regulate traffic anywhere except in Britain. He was a great admirer of the English way of coping with traffic and he wrote:

The English have always handled their street traffic with a dignity, courtesy and effectiveness which is a valuable lesson to all countries (Eno 1929: 181).

Though legal power to regulate traffic was given to the police in 1839 it took until the 1860s for traffic regulations to become part of police duties (Tripp 1950: 11). Some traffic signs were introduced with the Motor Car Act of 1903, such as speed limit and street hazard signs (Der Bundesminister für Verkehr 1984: 15–16). In 1910 the first international agreement on traffic signs came into force in order to unify the then existing different signs which varied between towns (ibid., p. 18). The London Traffic Act of 1924 and the Road Traffic Act of 1930 also included different traffic regulations, such as those for one-way streets which were introduced in London in 1924. Britain had a long tradition of pedestrian refuges which were probably introduced in London as early as 1870 and were also known from Paris (Eno 1929: 111). Marked pedestrian crossings were introduced in Paris in 1927. In London and in other British towns pedestrian crossings were installed largely as a result of the Road Traffic Act of 1934.

As in Britain, in Germany traffic regulations were controlled by the local police and were specific to each town. The first unified traffic signs which drew attention to dangerous conditions on the roads, such as train or street crossings, were from 1907 (Der Bundesminister für Verkehr 1984: 17). In 1909 the first Act to regulate road traffic was introduced and the acceptance of internationally agreed traffic signs started after 1910. A law to mark dangerous road conditions came into practice in 1923 and new international regulations in this matter were agreed in 1926 (ibid., p. 30). Many familiar traffic signs today were created during this period. Further agreement between the form, colours and style of traffic signs was achieved in 1931 but it took three years until they became part of the German street regulations.

However for many planners traffic regulations were not seen as being sufficient to cope with the future demand in traffic. Completely new ideas had to be developed, most of which were too utopian to become reality.

Separation of traffic modes – utopian plans for large cities

The first plans to construct London's underground in 1837 had stimulated utopian and grand ideas. Not only underground railways but particularly oversized arcades were seen as a way forward to ease traffic congestion. The building of London's Metropolitan underground line, which opened in 1863, had the purpose of relieving street congestion. Though the Metropolitan line was a success operationally its effect on traffic congestion was less significant.

These utopian arcades would have protected the pedestrians and would have acted as a connecting link either between the busiest parts of London or between railway stations. Two of the earliest projects were the Gye's Arcade, a glass-covered street connecting the Bank of England with Fleet Street, the Strand and Trafalgar Square, suggested in 1845; and the famous Crystal Way, discussed 10 years later. The latter was supposed to consist of a 4km long arcade between the City and the West End. The grandest project by far was Paxton's Great Victorian Way, a 'ring arcade road', 16km long, 22m wide and 33m high, connecting all the main railway stations with Piccadilly Circus (Choay 1969: 25). Most of these utopian projects included separate pedestrian areas. The main problem of these plans, all of which were enthusiastically discussed, was the cost. The cost of the Crystal Way was estimated at £2 million and the Great Victorian Way was calculated at £34 million, truly a sum far too high to be a realistic proposal. Ebenezer Howard's 'crystal palace' which was the planned shopping centre for his ideal Garden City, was an adaptation of Paxton's Crystal Way.

Other suggestions of multi-level segregation were made by the Royal Commission in 1905 when introducing the plan for two major avenues for Central London but this idea was also not put into practice (Royal Commission 1905).

Apart from British ideas to ease traffic congestion by separating traffic in the form of grand arcades, the *French* developed even more seriously the idea of separating different traffic modes in urban areas. Henry Jules Borie suggested in his essay 'Aerodomes: essai sur un nouveau mode d'habitations applicable aux quartiers le plus movementes des grandes villes' in 1865, to build large streets for pedestrians on interrelated terraces 20–30m high. Moving rooms which were driven by steam-powered elevators would have given people access to different street levels (Choay 1969: 20).

About fifty years later, Eugène Hénard expressed similar thoughts at the Town Planning Conference in London in 1910. Above the existing streets second streets would be constructed used by light vehicular traffic and pedestrians (Hénard 1910: 361). According to the level of traffic flows it would be possible to have three or four superimposed platforms. The highest platform would be for pedestrians and carriages, the second for tramways, the third for various mains and pipes and a fourth for the transport of goods. Obviously the third and the fourth platform would be underground. He concluded:

We should thus have a many storeyed street, as we have a many storeyed house; and the general problem of traffic could be solved, however heavy it might be (ibid., p. 362).

This French idea of separating transport modes into different levels or channels was expressed by the French repeatedly in the coming decades, in particular during and after the 1920s. Le Corbusier suggested, for instance in *Urbanisme*, streets which would be built on three different floors – two above and one below ground. Each street 'floor' would contain a specific transport mode. The 'ground-floor street' was to be used by pedestrians and non-commercial traffic. The first floor would cater for fast, one-way traffic on streets 120–180 yards wide (Le Corbusier 1924). In his later publication 'Manière de penser l'urbanisme' he further explained his idea of the vertical city. In this publication he classified five transport routes, of which two were for pedestrians, one for 'mixed use' – allowing a mixture of pedestrians and cars – but cars were obliged to drive slowly. However Le Corbusier had little time for the needs of the pedestrians. His cities were places for the motorists who could drive at high speed and not be held up by any traffic congestion.

The French architect H. Descamps designed his 'Cité moderne' (published 1927 and 1928) as a new town consisting of a network of 100m wide streets which were laid out in blocks of 500–1500m distance. Housing blocks were 20 floors high. The inside of these blocks contained large open areas between 40 and 90ha which were to be used as open spaces, play and sports grounds. Streets would be designed as one-way streets only, and footways for pedestrians would be built one floor up along the houses. The raised pedestrian walkways would lead to the inside yards of the houses with walkways down to the ground. A similar principle would apply for the town centre. According to this plan, pedestrians had little direct contact with motor traffic (Stübben 1929: 86).

The German Theodor Fritsch, who developed a vision of an ideal city which was in part similar to Ebenezer Howard's Garden City, suggested too the division of traffic into two levels. The surface level would cater for the transport of people, and the underground for transport of goods (Fritsch 1912: 20). During the 1920s several German planners believed in the idea of elevated footways, such as Scheibe (1925), Hilberseimer (1929) and Hegemann (1930) as will be explained in more detail in Chapter 5.

In the United States ideas of different vertical levels of transport were published. In 1897, William Eno suggested a four-level rail system, with a fifth level on top for bicycles for the City of New York (Eno 1929: 131).

One of the weirdest fantasies was *Roadtown*, published by Edgar Chambless in 1910. His idea was to lay modern skyscrapers on their side and design linear towns. The length of these towns could be anything from half a mile to hundreds of miles (Chambless 1910: 158). They were a combination between transport facilities and two storey-high apartments. All transport modes, apart from bicycles, would be electrical monorails. Roadtowns would not have streets or cars because none was needed. The transport system would be partly underground, consisting of express and local monorail trains. Two upper storeys would be built on top of the roadways, as apartments, and the centre of the roof would be used as a continuous glass-covered pedestrian promenade. On the outer edges there were paths for bicycles and even for roller skaters. Chambless concluded:

> The splendid view to be obtained from such a promenade in a dust free and smoke free country can hardly be pictured would apply for the town centre. According to this plan, pedestrians had little direct contact with motor traffic (ibid., pp. 53–4).

The Town Planning Exhibition in Gothenburg, Sweden in 1923 showed other examples of using different levels to cope with increased traffic flows. Corbett presented sketches of raised pavements or traffic on three levels (subway: heavy motor traffic; street level: vehicle traffic; raised pavements: pedestrians) for New York (Hegemann 1925: 50–1). Some German architects talked about the plans of two- and three-level roads in urban areas at the same time in which all the pedestrian traffic was to be raised to the upper floors (Wolf 1928: 870). The idea was rediscovered again much later in Britain by Colin Buchanan in his famous publication 'Traffic in Towns' in 1963 (Ministry of Transport 1963).

Utopian plans had little impact on the realities of urban life. They were exercises in intellectual fantasy. However their main value was that they expressed the need for fundamental change in order to be able to cope with existing and future traffic. One way was better overall planning which would combine the issue of transport with other functions cities and towns had to fulfil. Gradually at the beginning of the twentieth century, urban planning became an important concept and embedded in it were 'new' ideas not only on how to improve transport but also increasingly on how to improve road safety in built-up areas.

Urban planning as a new concept for controlling traffic

At the turn of the twentieth century, the idea emerged that cities should be planned and controlled in order to avoid the most negative developments which had plagued the cities as the period of primary industrialization became established. Urban planning as a municipal task included questions of road planning. Town planning in Germany, Britain and the United States developed at a different pace and with different emphasis. Whereas in Germany the town planning tradition started merely as street planning, which can be traced back to the late seventeenth century, in Britain the main town planning attempts were geared to improving health conditions. The United States had the weakest town-planning legislation though some rather radical approaches towards improving and beautifying some cities were made during the first decade of the twentieth century (City Beautiful and City Practical Movement). During this period there was a strong New-Baroque influence – Beaux Arts – in street design in the United States created by people like Burnham and Bennett.

By the end of the nineteenth century, attempts at controlling traffic by way of developing a street hierarchy were developed in the most sophisticated way by German engineers and planners. Baumeister's book *Urban Extensions* (*Stadterweiterungen*, 1876) referred to the need for good road planning to consist of a main street network for all kinds of traffic and side streets for residential traffic. He argued that in some circumstances three or four different street classifications were needed (Baumeister 1876: 95). In practice, many German cities had already adopted a variety of different street widths which were used according to the road function assumed when new urban areas were planned and constructed.

Another German, Theodor Goecke suggested in 1893 a stricter functional division between residential and traffic streets. The traffic street was the surrounding street for a large housing block, and the residential streets would give access inside them. His idea of the residential street was derived from the medieval lane which would largely protect residents from wheeled traffic. He argued against the formal division between traffic and residential streets which fulfilled in his opinion no purpose, because so-called residential streets could easily be converted into traffic streets (Goecke 1915 and 1918).

In Britain the approach towards roads was totally different to Germany. It was derived from health arguments. Wide streets would enable sufficient air circulation between houses. Light and sunshine had been discovered as an important factor in controlling some serious illnesses, and therefore became the determinate factor for street layouts, after local authorities had gained the power to control street designs during the last quarter of the nineteenth century. With the development of urban planning as a serious subject, traffic considerations became an issue for Britain too. New ideas on street layouts were developed mainly in the newly built garden cities and garden suburbs and in residential areas which were influenced by the garden city movement or its intellectual predecessors.

In the United States the grid-iron street layout was primarily used for towns and town extensions (Hegemann 1925: 11). Often the grid-iron was complemented by relatively wide and long diagonal streets. Normally the street widths would be rather uniform, and only the main streets were somewhat wider. There were some exceptions, especially the streets designed by Olmsted which showed not only a greater variation but also abandoned the grid-iron network.

The earliest pedestrianization schemes and other forms of traffic restriction

During the past centuries, a multiplicity of building forms had been developed in most towns and cities throughout Europe which protected pedestrians from wheeled traffic, such as open passages, courts, through houses, lanes or simply private streets. But now with the increase in car use and the 'murderous invasion' of the motor bus (Webb and Webb 1913: 254) an adaptation of the street network and road surface to these new transport modes was demanded. The most important question which evolved because of the constant increase in traffic volumes and the different demands on the streets was, as the Webbs put it:

whether traffic was to be restrained to suit the road, or the road constructed to accommodate the traffic (ibid., p. 242).

Early forms of pedestrianization – road closures and other restrictions on vehicle traffic – may have been a feature in many towns in which the streets were too narrow to accommodate wheeled traffic. It was probably so common that in most cases there are no documents mentioning it. In a country like Germany which had very strict traffic regulations given by police order, documents are more easily available. But even in Germany we have relatively little evidence and if so then only brief accounts. Police orders for German cities showed that traffic restrictions or the exclusion of different forms of traffic were quite common (for further details see Chapters 3 and 4).

In Britain, early examples of road closures in urban areas are also difficult to find. One of the reasons may have been that in general the attitude towards traffic was more relaxed and it may not have been as easy to restrict traffic. However Olmsted pointed out that during the nineteenth century in certain streets of London, transport of merchandise was not allowed during most hours of the day (Sutton 1971: 23). In general, improvements for pedestrians came from bodies concerned to provide sufficient open space and to preserve public footpaths, such as some of the City corporations. An interesting example of this attempt was provided in Leicester, the 'New Walk', which is referred to as a pedestrian road. The 'New Walk', designed at the end of the eighteenth century, was one mile long and people could walk from the suburbs almost to the centre of the town. Though the walk was quite wide (nearly 9m) no wheeled traffic was allowed.

In the United States ideas to exclude some traffic were common. Burnham wrote in his *Report on a Plan for San Francisco* that along the boulevards heavy traffic should be restricted and on some streets not allowed at all (Burnham 1906). In Chicago along the main boulevards commercial and other wheeled traffic was restricted on most weekdays up to 1920. Hegemann and Stein remarked that some North American cities closed city centre streets to motor traffic in the 1920s (Hegemann 1933; Stein 1925). Ford was of the opinion that the essential condition of good planning was the provision of a large space which could be kept completely free of wheeled traffic or from which traffic could be excluded (Ford 1920).

Another interesting example of an early pedestrianization scheme can be found in Buenos Aires, Argentina, in 1911. According to Loew, the Municipality of Buenos Aires forbade traffic in the most prestigious street in the town centre from 11.00 to 21.00. It was simply a recognition of the fact that the citizens had taken the street (the Florida) as their favourite promenade. The Florida has been closed to traffic since then and was fully pedestrianized in 1971.

The idea of restraining traffic was not uncommon in France. Le Corbusier complained:

Table 2.1 Inhabitants per
motor vehicle

	1922	*1938*
USA	10	4
Great Britain	91	19
Denmark	131	26
France	176	20
Switzerland	219	–
Belgium	228	37
Germany	360	44

Sources: Ehlgötz (1925: 45);
Verkehrstechnik (1939: 199).

'So much the better' said a great authority to me, one of those who direct and elaborate the plans for the extension of Paris, 'motors will be completely held up' (Le Corbusier 1924: 16).

By the second decade of the twentieth century the use of motor vehicles had further increased, though the degree of motorization varied tremendously between the different countries, as can be seen in the next section.

The degree of motorization in some European countries and in the United States

Before World War I cars in Europe were mainly built as luxury and sports cars, and became more important for the economy during and after World War I. Table 2.1 shows the degree of motorization in some European countries and in the United States in 1922 and 1938. The United States had by far the highest car ownership level, but Britain was the most motorized country in Europe, and Germany lagged far behind.

In Germany, there were substantial differences in the degree of motorization between the large cities and the countryside, and also between cities, e.g. Berlin counted one motor vehicle per 20 inhabitants, Munich 13, Cologne 20 and Essen 34 in 1938 (Verkehrstechnik 1939: 197). Clearly this was also true in Britain. Unfortunately there are no data easily available for the same year and for individual towns as in Germany.

From Table 2.1 it is obvious that Germany must have had fewer problems with motor traffic than Britain. Furthermore Germany introduced a comprehensive driving test already in 1909 while in Britain it took until 1934. In both countries there was great concern expressed about the increasing number of accidents in which motor vehicles were involved and the negative impact that cars and motorization had on the environment. In addition, the Germans strongly emphasized the negative effects that motorization could have on the urban structure, especially the city centres. This concern was partly caused by traffic consideration and safety aspects, but also by the feared destruction motor traffic would have on the 'aesthetics' of cities. This will be explained in more detail in Chapter 3.

Conclusion

Studying history in this way one is astonished by how many different devices and ideas were used to cope with wheeled and later motor traffic. The most common form throughout consisted of suggesting or actually planning the separation of the different transport modes. This was carried out in the form of pavements, or in more sophisticated ways such as arcades. Upper-level walkways were frequently suggested but were far too expensive to be realistic.

One of the new forms of protecting pedestrians in the city centres was the arcade, built primarily during the nineteenth century. Although arcades were very effective, they were too expensive to be used in any more comprehensive forms, although there were several such suggestions.

Traffic separation was used in parks and one could argue that these were the models for separating pedestrians from motor vehicles later applied to some extent in Britain and in the United States for residential areas.

The 'art' of regulating traffic was very advanced in London, which is not surprising because of the severity of its traffic congestion. Indeed, all countries which were by the standards of the day highly motorized very quickly adopted a whole range of traffic regulations.

It is particularly interesting that the 'leading' nation in utopian planning was France. Again one could see some links between these ideas and many of the projects which have recently been built in Paris under the Mitterrand presidency.

The differences in the political background and attitudes between Germany and Britain by the end of the nineteenth and the beginning of the twentieth century led to a different approach towards motorization which will be discussed in the following chapters. But we can already see an advance in street planning and a tendency to a greater readiness to close streets in city centres in Germany than in Britain. The concern for pedestrian rights came from the 'environmental lobby' in Britain unlike Germany where, as will be seen later, it came primarily from the national heritage movement.

3 Street planning in Germany and Britain from the nineteenth century to the beginning of the twentieth century and the appearance of the motor vehicle

Origins of street planning in Germany: the nineteenth century

This chapter could be regarded as crucial in the attempt to unravel the major elements of the difference between street planning in Germany and Britain. In both countries it was the demand for wheeled traffic which changed street layouts from the sixteenth to the eighteenth century. It is primarily in Germany that we see numerous examples, during the Baroque era, of town extensions. Although there are fine examples of town extensions in Britain during the eighteenth century, they were not very common.

German towns and cities were further extended during the nineteenth century and street layouts were still derived from Baroque street designs. Avenues and the large squares were lavishly used. Again we do not find this to be the major characteristic of the British cities during the same time period.

As a result of industrialization and population growth, towns and cities grew at a rate not experienced before. In Germany most of the larger town extensions took place just before or after the formation of the German empire in January 1871. Yet all these later town extensions were modelled on Paris, Vienna and Berlin. Let us briefly look at the street designs in these capitals.

In 1853, the Alsacian Georges-Eugène Haussmann became responsible for substantial urban improvements in Paris which consisted mainly of planning new boulevards, squares and parks. The new plan had to fulfil two major objectives. Paris was intended to become the most beautiful city in the world and the new plan was seen as the artistic expression of Napoleon's III new empire. The second objective was to satisfy military needs which implied the removal of many narrow and crooked streets and complex corners which were so easily transformed into barricades. Streets designed according to military purposes were not a particularly new idea. It was first expressed by Leone Battista Alberti during the fifteenth century. He divided streets into military and non-military streets.

The Haussmann plan, which was regarded as a masterpiece of street planning at the time, resulted in the demolition of many historic quarters. It was nearly complete by 1870, but further improvements took place between 1871 and 1898. This plan brought major improvements for pedestrians. Along the wide boulevards of which 95km were wider than 24m, generous pedestrian pavements were built, bordered by trees. Pavements were extended from 141km to 1,290km and 60ha of new parks and squares complemented the street planning.

Vienna had developed a town extension plan removing the historic city wall and replacing it with a 4km long and 57m wide ring road which had the character of a promenade in 1858. Many of the most impressive buildings of the Habsburg empire were to be built there. In addition many new boulevards were laid out.

Not surprisingly, Berlin could not be left behind after the Emperors Napoleon III and Franz Joseph had beautified their capitals. In 1861, James Hobrecht was appointed to draw up a street plan for the future city extensions. A network of wide boulevards was planned including a ring road similar to Paris and Vienna. About two decades earlier, Peter Joseph Lenné, the garden architect of Friedrich Wilhelm IV, had designed a ring road intersected by large squares. Some of these squares became later part of the Hobrecht plan. But in contrast to Paris the historic city centre of Berlin stayed largely untouched. Even so, the city centre changed, including the demolition of historic buildings, in its socio-economic function and became increasingly the centre of the German Reich.

There has been a significant amount of criticism of the Hobrecht plan. According to Matzerath smaller streets were supposed to open up the large housing blocks (up to 400sq.m) (Matzerath 1984) but this never materialized. Hegemann and others were of the opinion that the plan had increased land speculation (Hegemann 1930). Hartmann used an example of the estimated increase in the value of building land in Berlin. A building plot which cost 100,000 RM in 1860 was worth 300 million RM only thirty-seven years later (Hartmann 1976). Edward Unwin noticed that suburban land in Berlin was 3–6 times more expensive than land in a similar location in London (E. Unwin 1931).

Many other German towns designed town extension plans which were modelled on Paris, Vienna or Berlin. Some of these extensions were already part of international competitions, such as the town extension plan of Cologne 1880 and of Munich 1892–3. As in Vienna many German towns removed their historic walls and changed them into wide ring roads, e.g. Hamburg, Munich, Leipzig, Breslau, Bremen, Hanover.

The ideological background: Reinhard Baumeister, Hermann-Joseph Stübben and Camillo Sitte

A new more sophisticated concept for town extensions was propagated by Reinhard Baumeister. His book, *Stadterweiterungen in technischer, baupolizeilicher and wirtschaftlicher Beziehung* (Town Extensions in their Technical, Surveying and Economic Relationship), was published in 1876. It was extremely influential in setting new standards for street and urban planning. In terms of street planning his suggestions were far more advanced than practised. In his opinion, through traffic should be taken out of the built-up areas (Baumeister 1876: 38) and the width of the streets should be built according to traffic volumes. He actually calculated the capacities of streets and set upper limits (he used the number of pedestrians/vehicles per metre of pavement/carriageway and divided them by time units). Clearly the upper limit differed according to traffic speed and transport mode (and comfort).

In Baumeister's opinion a good street network should consist of main roads and subsidiary streets. The smallest subsidiary roads should not be less than 8m, including pavements (ibid., p. 118). Roads carrying more traffic should be 17m and main streets 25m wide (ibid., p. 119). As a rule the ratio between major carriageway and pavements was calculated $\frac{3}{5}:\frac{1}{5}$, leaving $\frac{1}{5}$ for trees.

Whereas the main streets would cater for all kinds of traffic, and would consist of different lanes for parked, slow and fast traffic; the subsidiary streets would allow only residential traffic. If the traffic flow was very heavy, vehicle traffic would be divided into separate carriageways in order to make it safer for pedestrians. Such streets should have a middle promenade for pedestrians, measuring about 9m. For even wider streets he suggested three carriageways and five pavements in succession. Because of his concern about pedestrian safety he was in favour of a large number of narrower streets rather than a few wide ones. Two parallel streets of 15m were better than one of 30m.

In most circumstances one would need three or four different classes of streets. He gave examples of different actual road widths in German cities. In most of them at least four types of street width were already in practice. There should be lively and quiet roads according to the housing needs. His ideal overall street layout had the form of a triangle. The knots of the triangle consisted of major traffic generating points, such as bridges, stations, squares or gates. Both the major and minor street network were made up of different triangles which set together would form rectangles. The main arterial streets were the diagonals of the rectangles. This street network was complemented by ring roads. Baumeister was of the opinion that the more ring roads the better.

Street layouts which were planned in the form of squares or rectangles – without the diagonals – were seen as boring. He had strong feelings against curved roads which were incompatible with the new demands of traffic, though he thought that curved roads would work if there was natural unevenness. But artificially curved roads were unrealistic in modern town extension plans. In his view one could not artificially recreate history.

If there was too much traffic in the city centre, then streets had either to be widened or, even better, parallel streets built. A new street network had to be laid above the old one. He was not in favour of removing medieval walls but he advocated the removal of more recent military walls; the space was to be used for ring roads.

In fact, Baumeister's book was very advanced for its time. It was in a sense a technical handbook on how to plan better town extensions but it was more than that. One of the crucial elements was the division of urban space into industrial, housing and commercial quarters.

Baumeister's concept was seen as the ideal concept to cope with the modern demands of traffic. It was therefore very influential, and most towns had at least plans to use his principles, e.g. the Gnauth plan in Nuremberg of 1876 or the Lasne Plan in Regensburg of 1917.

Hermann-Joseph Stübben had very similar ideas to Baumeister. He had been responsible for planning in Cologne and had been in favour of a wide 6km-long ring road around the historic centre of Cologne (Wurzer 1974: 18). He became well-known after his book *Der Städtebau* was published in 1890. Stübben criticized the Prussian Building Laws (Fluchtliniengesetz of 1875) which were used for Prussian town extensions because the classification of streets was too coarse. Major traffic streets had a width from 30m upwards, traffic streets measured between 20m and 30m and residential street widths were between 12m and 20m (Stübben 1890: 67). Stübben was of the opinion that much smaller streets should be used if there was a low traffic flow. But he was clearly against too narrow streets in densely built-up areas, they should be widened to 6–7m. He was in favour of different street widths according to the demands of the traffic.

Stübben's book referred to the town extension of Cologne and mentioned the different street widths used. The plan showed streets which had widths of 12, 14, 16, 18, 20, 22, 26 and 30m; in addition the ring roads were between 32m and 70m wide. He already warned against building streets too wide because houses would be built too high. This was a lesson learned from the Hobrecht Plan in Berlin.

Like Baumeister, he was against curved streets because they produced disadvantages for traffic. But straight streets should not be longer than 1km, otherwise the street would become too boring. He admitted that winding streets were more interesting and should also be part of modern town plans. However, to have curved streets as a rule of modern town planning, or to copy the old historic street layout, was foolishness.

Baumeister and Stübben both had similar ideas on what was necessary to cope with the increasing demands of town traffic. It was obvious that the medieval street layout which most German towns had was not very well suited for this purpose. Baumeister was more radical in his approach on how to deal with town centres, whereas Stübben tended to be more in favour of restricting traffic in the centre and even closing streets to vehicle traffic if necessary.

Let us pause here briefly and recapitulate the political significance of the type of planning which was practised. As discussed above the existing urban planning was a continuation of Baroque style. Baroque as an art element was largely regarded by the Germans as foreign (despite plenty of Baroque German churches). It was the art of absolutism, a symbol of repression. But, what was even worse, this type of town planning had its most recent expression in Paris, the capital of the country Germany had just fought a war with in 1870–1. The proclamation of a unified German Reich had taken place in the palace of Versailles after a stunning but bloody victory over the French army. It was widely acknowledged that the Hobrecht plan was a vague copy of the Haussmann plan. It was also an expression of Prussia's imperial power. The newly formed Prussian Empire was itself riddled with social tensions which did not only affect the working class but there were many liberal and conservative groups which had difficulty in identifying themselves with the new government.

Though there was hidden and open criticism against the Empire and its failure to allow social reforms and political freedom, this did not automatically affect patriotic feelings. There was a dangerous mixture between romanticism and patriotism and it included both the left- and the right-wing parties. This newly formed German Empire, though a real creation, was for many Germans some kind of romantic rebirth of the medieval Holy Roman Empire of the German Nation (Heilige Römische Reich Deutscher Nation), which had been formed by the German King Otto I in the year 962.

Patriotism and romanticism influenced art and literature. Medieval art styles became fashionable and it also determined how cities were perceived. The medieval city with its crooked narrow streets, its casually formed squares and its timber-framed houses, needed to be preserved, and such street layouts were seen as something positive, as German culture *per se*. The Fuggerei in Augsburg, the first social housing colony built for the old and poor by the most powerful banking family of Germany, the Fuggers, in 1519; or the large medieval courtyards (*Wohnhofarkaden*) became models of housing. Typical were sheltered lanes inside the colony (Fuggerei) or the large housing blocks.

But there was Baumeister's book, which taught German officials that these cities did not work any more and that modern traffic demanded the partial destruction of crucial elements of these cities, the city centres for instance. I think we cannot imagine what the new publication by Camillo Sitte, which came on the market in 1889, meant for many Germans. It was, among other things, a polemic criticism against Baumeister's famous publications.

The book *Der Städtebau nach seinen künstlerischen Grundsätzen* (City Planning according to Artistic Principles) was so popular that after 6 weeks a second edition had to be printed and only one year later the third edition appeared. In 1902 the book was translated into French though the French hardly acknowledged it, probably because it was foreign to the French understanding of cities and maybe because it was too 'German' (Camillo Sitte was actually Austrian). The fourth edition was printed in 1908 and included a new chapter on open space in large cities (Großstadt-Grün).

Sitte's view of urban planning differed strongly from the ideas of Baumeister and Stübben. He despised their approach completely and argued that:

City planning should not be merely a technical matter, but should in the truest and most elevated sense be an artistic enterprise (Sitte 1965: 4).

He developed ideas of artistic principles in urban planning which as he pointed out existed in past centuries and compared these with the 'modern' – nineteenth-century – approach to urban planning. His main concern was the historic street layout, in particular squares, of traditional towns and cities. The idea that town squares were of primary importance to the life of every city, or had (in the Middle Ages and the Renaissance) a vital and functional use for community life, was well understood by professionals and politicians. In his opinion:

Major plazas and thoroughfares should wear the 'Sunday best' in order to be a pride and joy to the inhabitants to awaken civic spirit (ibid., p. 92).

Sitte was against street widening because in his opinion it very often had nothing to do with traffic flows but was only a fashionable thing to do. The ideal street was picturesque, curved instead of being straight, and had to

form a completely enclosed unit . . . Moreover, the winding character of ancient streets kept sealing off perspective views in them while offering the eye a new aspect at each succeeding turn . . . (ibid., p. 61)

The desire to set off a well designed building or a natural feature is surely the reason for concave curves in old streets . . . (ibid., p. 65).

Sitte's attitude was the most marked division between a technical and an artistic-historic approach towards the planning and building of urban streets. This division would become a constant conflict between planners for the coming decades. His ideas stood in contrast to a purely technical and functional approach to street design. He wanted to preserve the historic cities, and with them the historic squares, thoroughfares, streets and lanes. When, in the 1920s and later on, the concern about motor traffic evolved, it was Sitte's inheritance which made planners aware of the problems. Sitte had many supporters and his book had already become a classic in the 1920s. (In 1921 a fifth edition was published.)

It was above all the division between the technical and aesthetic approach to street planning which was totally lacking in Britain. Britain not only had little expertise in street planning, as will be discussed in the next section, but also it did not go through a period equivalent to the glorification of the medieval city centre so crucial to Germans.

Street regulation in towns during the nineteenth and early twentieth century in Britain

In Britain, roads, byways and country lanes had been for centuries under the jurisdiction of the parish surveyor of highways (Webb and Webb 1913: 193). By the start of the nineteenth century over 1,000 Turnpike Trusts had been set up to assure a reasonable road network and to carry out necessary road repairs. Apart from the main inter-urban roads the Turnpike Trusts also looked after some of the thoroughfares. But even in the first half of the nineteenth century, in 1838, the Trusts took care of just over one-sixth of all roads in Britain.

The majority of roads were under the control of at least 15,000 separate highway parishes in England and Wales alone. Obviously, the condition of roads was dependent on the priority a parish gave to them. The Webbs pointed out that during the first thirty years of the nineteenth century the same highway mechanism took place as during the sixteenth and seventeenth centuries.

The Public Health Act of 1848 set up a Central Board of Health which in turn created local boards of health (Ashworth 1965: 58). The Surveyor of Highways came under their jurisdiction. The Local Government Act of 1858 increased the powers of local boards of health but there was still significant confusion over which local body was responsible for what function in a town. There were the borough councillors who could make bye-laws, the highway boards which were created under the 1862 Highway Act and there were sanitary authorities. All three bodies could well be partly responsible for the condition of roads. However, which body was responsible varied enormously from one local authority to another.

The Public Health Act of 1875 divided the country into urban and rural sanitary districts. These districts also kept the power over their highways. By the end of the nineteenth century most highway districts had largely been replaced by urban and rural sanitary authorities. The same Act also changed the jurisdiction of highways and turnpike roads at the highest level. The newly created Local Government Board took over this responsibility from the Home Office.

The Public Health Act enabled urban sanitary authorities to control certain features, such as street widths, building heights etc. However such regulations (bye-laws) had already been in existence since the middle of the nineteenth century in some cities through the authorities of local Acts in Parliament. London for instance had exercised bye-laws since the setting up of the Metropolitan Board of Works in 1855.

The most severe problem with bye-laws was that most of the regulations were designed by bureaucrats who knew little if anything about the technical, transport or planning aspects of streets. Unwin pointed out that the bye-laws were set up under the consideration 'of the man who has to administer them, and not of the man who has to work under them' (Unwin, 1971: 390).

Section 4 of the Public Health Act specified that basically anything 'in the way of road or path' was defined to be a street. According to bye-laws, the enforced width of streets was not determined by existing or expected traffic flows, building heights or the characteristics of the area. There was no difference in these regulations for the city centre or a newly designed housing area on the outskirts. In the back of the minds of the designers were the conditions of housing developments of the early industrial era with narrow roads, culs-de-sac, historic courts and back alleys which had become squalid and filthy slums. Charles Dickens in his novel *Hard Times* or George Gissing in *The Nether World* had described such streets vividly.

To walk about a neighbourhood such as this is the dreariest exercise to which a man can betake himself; the heart is crushed by uniformity of decent squalor; one remembers that each of these dead faced houses, often each separate blind window, represents a 'home', and the associations of the word whisper blank despair' (Gissing 1889, quoted in Williams 1975: 282).

Sufficient sunshine and air could not penetrate in the lower parts or rear sides of such housing, which in turn was partly responsible for epidemic illnesses, such as tuberculosis. It was the health aspect alone which favoured sufficiently wide streets. The street widths required differed according to the local authority. A 1862 Act in London demanded that newly built streets had to be at least as wide as the height of the buildings. But this restriction applied only to streets under 15.25m (50ft). As soon as a street measured 50ft or more, houses could be built as high as 24.4m (80ft) (Muthesius 1979: 75).

Other regulations showed even less understanding about the functions of urban roads, e.g. the regulation that a street had to be 40–50ft (12m–15m) wide if the road was longer than 100–150ft (30m–45m) (Unwin 1971: 391). In some authorities, cul-de-sac roads were not allowed. Another common regulation was that after a particular length of street, crossroads had to be built. Robinson referred to the bye-laws of Liverpool in which streets longer than 150 yards had to have a cross street (Robinson 1916) which made the whole street layout extremely boring and monotonous.

Bye-laws were only applicable within the boundary of urban sanitary districts but not in the rural equivalent (Ashworth 1965). Therefore any suburban development outside the border of urban sanitary districts was out of the control of local authorities. This aspect was important for the development of the model villages by industrialists, such as Cadbury (Bournville), Lever (Port Sunlight) and Rowntree (New Earswick) which were outside the control of urban sanitary districts. As bye-laws were not applicable in these villages, architects could develop a much more imaginative street layout. In New Earswick and Letchworth no bye-laws were in force.

A relaxation of bye-laws was brought about firstly with the Hampstead Garden Suburb Act in 1906 which allowed the use of culs-de-sac and more informal street planning in general. Hampstead Garden Suburb was located in the urban sanitary district of Hendon and therefore controlled by bye-laws. This Act suspended the local bye-laws for a minimum road width. It was the predecessor of the 1909 Town Planning Act which gave local authorities greater flexibility in several aspects of planning, but also limited the density, height and character of buildings (Creese 1967: 21).

However, one of the major problems was still that even in the first quarter of the twentieth century English highway administration 'was still split up among nearly 1900 separate local authorities' (Webb and Webb 1913: 243) and there was an increase in this number and no central control over them. Finally the 1929 Local Government Act made county councils responsible for all roads outside the major urban areas.

British and German street planning in contrast

The German planning professor, Berlepsch-Valendás who visited England regularly from about 1900 onwards, made some outspoken remarks about the English street regulations. He referred to the Garden City of Ealing which to his knowledge could not be developed because the local authorities insisted on 'almost nonsensical street width'. Only a parliamentary order solved the problem (Berlepsch-Valendás 1913: 85). He also criticized the street width in Woodland, a colliery village near Doncaster which contained streets which were 20m (66ft) wide excluding footpaths, front gardens and lawn borders. The one-and-a-half-storey-high houses disappeared completely behind such street widths. The wide streets created such dust that they reminded him of a sandstorm in the Sahara when the wind was blowing. He concluded that for local authorities these considerations did not matter as long as the orders were obeyed.

One of the most interesting publications which pointed out the differences between Britain and Germany in terms of street layout was Hermann Muthesius' publication: *The English House* which had a great influence on German house design. Hermann Muthesius, a German diplomat but an architect by profession, lived and worked from 1896 to 1903 in Britain. During his time in England he wrote most of his three volumes of *The English House* which were published in Berlin in 1904 and 1905. Though Muthesius' main interest was to study the different forms and styles of contemporary English houses, he also included some remarks about the street layout in Britain.

Residential streets cross one another in wild confusion, without any kind of guiding principle. Only the initiate can find his way through this remarkable system. A few thoroughfares that were formerly the old high roads provide the only guide. These soon become overloaded with traffic, a condition that is typical of the narrow old streets of London. Traffic can not be diverted into a parallel side street because none exists. . . . Probably no area today provides a better idea of the medieval city than the network of streets in which the Englishman of the twentieth century lives. The cheerfulness with which he submits to the intelligence of the speculative builder in this matter is amazing (Muthesius 1979: 140).

Muthesius pointed out that in contrast to Germany local authorities in England were not concerned with the street layout. Each developer or ground landlord can plan the streets to suit his own requirements. Local authorities had very few rights to interfere in the street layout. Only some regulations about the width of streets had to be fulfilled. Permission could be refused if a carriageway was less than 12.20m (40ft) and pedestrian walks less than 6.10m (20ft), if culs-de-sac were longer than 18.3m (60ft) and carriageways steeper than 1:20. In certain cases the pedestrian walks had to be converted into carriageways and main streets widened to 18.3m (60ft). Muthesius concluded that the continental procedure to have streets planned by local authorities had some 'darker' aspects but the English procedure of caring not at all about these important matters was clearly an immense defect and an incomprehensible sign of backwardness. He concluded 'the planning of the whole city is thus handed over to the lowest order of intelligence' (ibid., p. 137).

However there was one form of development in Britain which was keen on experimenting with new street layouts and soon became a model for many German planners. These were the low-cost industrial housing villages, later the garden cities and suburbs which will be discussed in Chapter 4.

The deadly motor car in Britain and Germany

According to the Webbs the first motor cars were introduced to Britain in 1894 (Webb and Webb 1913: 240). The 'Red Flag' Act – properly called the Locomotive Act – implemented in 1865, required that a 'self-propelled' vehicle was not allowed to drive faster than 4mph in the countryside and 2mph in towns, and had to be preceded by a man waving a red flag (Plowden 1971: 22). This regulation was repealed by the 1896 Locomotives on Highways Act, and the speed was increased to 14mph on public roads, but lower speeds could be implemented by the local government board.

Germany certainly did not have anything like a 'Red Flag Act'. Regulation of speed, maintenance of roads and other questions of car use were in the hands of the different states (*Länder*), provinces or cities. For instance, road maintenance for the 63,000km of state or provincial streets in 1931 (out of a total of 274,000km) was carried out by the different states. However in Prussia the various provinces had this responsibility. For the majority of roads different secondary or local authorities were responsible (Wohl and Albitreccia 1935: 126). Speed limits were regulated by the different *Länder* which in turn allowed cities or provinces to make their own regulations. The example of Berlin and the Rhine Province showed variation between 12 and 15kph in built-up areas before 1909 (Horadam 1983: 19).

Apart from motor cars, the use of bicycles as a transport mode became increasingly common from 1900 onwards, in particular for the young and lower income groups in both countries (Webb and Webb 1913: 239). Yet cycling was more popular in Germany and already in 1914 there were 6 million cyclists. In 1935 the number had increased to 16 million while England had 8 million (Schacht 1938: 170). But one has to take into account the different population size, Germany had 67 million whereas England had only 37 million.

In addition, motor buses started to run just after 1900. The Webbs wrote that around that time people had to become accustomed to roads which belonged not only to the local residents but to everybody who could mount a bike or a motor vehicle (Webb and Webb 1913: 240). This was not quite accurate because the busy roads in the major cities had to cope with a significant amount of wheeled traffic during the late eighteenth and nineteenth centuries.

Again we see differences from Germany where the motor bus was not needed in large numbers because the electric trams had become the main urban transport mode and a large network was developed which would in many towns reach far into the hinterland. This development could not take place in Britain because of the more complicated legal requirements for trams. Local authorities leased the tracks to the operating companies for 21 years. According to Mackay and Cox, 21 years were normally too short a period to recoup the investment for electrification (MacKay and Cox, 1979).

Hamer described the closeness between the interests of cyclists and motorists in Britain. The two largest bicycle organizations – the Cyclists' Touring Club and the National Cyclists' Union – had formed a pressure group in 1886, called the Road Improvement Association (RIA) whose objective it was to improve the condition of roads which were largely unsuitable for cyclists and motorists alike. The Automobile Club – called since 1907 the Royal Automobile Club – was very keen to merge with the rather powerful Cyclists' Touring Club but was rejected. As a result the Automobile Club increased its influence in the RIA (Hamer 1987: 24).

With the increase in the number of cars, opposition against motorists began to grow around 1900. The dislike for cars, which Plowden described as hysteria (Plowden 1971), came primarily from the rural population because they saw their roads wrecked and had most to suffer. The dust and mud created by cars resulted in some cottages being almost uninhabitable and greatly depreciated in value (Webb and Webb 1913: 241). There was also growing opposition from other parts of the country due to reckless driving and the increasing number of accidents. The *Daily Telegraph* started an anti-motorist campaign, calling the motorists 'the Social Juggernauts'. There was a vigorous campaign by the police against motorist offenders and there was a considerable antipathy from various magistrates when those offenders came to trial.

There was an early awareness of the danger cars inflicted on society. This was partly the result of very primitive roads and the novelty of the mode but was primarily caused by the large number of accidents. The negative image cars had became part of British popular culture and can even today be sensed by the children's book *The Wind in the Willows*, first published in 1908 (Grahame 1908).

It is interesting to note that very similar sentiments against cars can be found in Germany during this time. In 1908 protests were expressed about car drivers in the Prussian Parliament:

Car drivers show very often a spectacular insensitivity . . . and the population is extremely bitter about them in particular because they flee if they have done any kind of damage (Sachs 1984: 24).

In 1912 a protest was published by Michael Freiherr von Pidoll. He asked:

From where does the car driver take the right to rule the street which does not belong to him but to everybody. But not only that, he also interferes with everybody on each step and dictates everybody's behaviour. The public street is not made for express traffic but is part of the urban environment . . . (ibid., p. 27).

Despite these protests by the population in both countries, the different groups promoting the motorists, such as the SMMT (Society of Motor Manufacturers and Traders), the RIA, the AA and the RAC in Britain, and the Benzolvereinigung, the Nationale Automobilgesellschaft and the Automobil- Verkehrs- und Übungsstraßen GMbH (all of them formed in 1909) in Germany made sure that changes according to their needs occurred. In both countries cars were expensive vehicles driven by a rich and politically powerful minority, although in Germany the membership of car clubs was very low. In 1902 the total membership was only 900 and increased only slowly (Bardou, Chanaron *et al.* 1982: 18) although the German Emperor had by then reversed his original opposition to the car.

In Britain the 'road lobby' pressed for a parliamentary change of the existing speed limit. It wanted the speed limit removed and fought against car registration and driving tests.

The 1903 Motor Car Act was a near disaster for the road lobby. A new speed limit of 20mph was implemented and gave the Local Government Board the power to impose 10mph at the request of local authorities (Plowden 1971: 57). The Act also introduced car registration but did not include driving tests. Several motoring journals thought the Act 'would cripple if not annihilate the motor industry' and that it was 'a disgrace to an engineering country'. In reality nothing of the kind took place.

High accident rates were blamed to a large extent by the road lobby on the appalling road conditions. It is further interesting to note how the road lobby in the coming years successfully reduced this bias against motor vehicles and was able to turn the tide in favour of them.

In Germany the first unified law with reference to motor vehicles was passed in 1909 (Gesetz über den Verkehr für Kraftfahrzeuge vom 3 Mai 1909) which was followed by the regulation for motor vehicles (Verordnung über den Verkehr mit Kraftfahrzeugen vom 3 Februar 1910). Among other things, it introduced the lengthy procedure of obtaining a driving licence and regulated the speed limit to 15kph (9mph) in built-up areas (Bongard and Bongard 1983: 10). It was increased in 1923 to 30kph (about 19mph) in built-up areas but local authorities could modify it up to 40kph. Berlin had for instance a speed limit of 35kph (Horadam 1983: 20). No specific speed limits were introduced outside towns and villages, drivers had to drive according to conditions of the roads.

In 1923, Germany had about the equivalent speed limit of Britain in 1903 which would roughly correspond with the delay of German motorization. The number of cars and buses used in Germany was much lower and partly as a result of that the pressure from the German road lobby was weaker and only increased during the Third Reich. Germany had in 1907 only a third of the number of cars as Britain (for more details see Table 3.1 and 3.2, pp. 43 and 44). Yet from 1925 onwards, the number of motor cycles increased rapidly and until 1939 varied between 46 and 48 per cent of the total number of motor vehicles.

The following section concentrates more on the events in Britain (more details on Germany are given in Chapters 4 and 6).

The Third International Road Congress in Britain

The importance of motorization was also expressed in the holding of international road congresses which, among other issues, tried to set standards of road design and developed ideas on the ideal street network. The First International Road Congress, which was held in Paris in 1908, was organized by the British Road Improvement Association (RIA). The Third International Road Congress, held in London in 1913, discussed the needs for main arterial roads which should either be widened or new ones built. The resolution of the Congress demanded the construction of bypass roads which were needed for towns of secondary importance located on main roads. It was not advised to widen existing roads passing through such towns. Adshead argued that the Congress resolution in favour of constructing bypasses was passed more with a view to the safety of the urban population than for the convenience of the motorist who quite liked to drive through towns and stop there (Adshead 1915). During this Congress, it was further agreed that standard main roads should have, apart from tram tracks, waiting, slow and fast lanes; 40ft (12m) was regarded as a minimum width for main roads. Martin pointed out that in the General Report by the Permanent International Association of Road Congresses in 1913, a classification of four road types was suggested:

- Class I should be called national or trunk roads; these were the main roads which would connect the principal cities and ports.
- Class II were roads which connected the trunk roads and also the towns and villages; they should be named country, or departmental roads.
- Class III were local roads, including streets in towns and roads in country districts; they could simply be referred to as local roads.
- Class IV were 'tracks, bridle paths, rights-of-way and other means of communication over which the public has the right to travel but which were not used by motor traffic' (Martin 1915: 63).

It appears that at that time there were still some close links between engineering and town planning. Engineers were not yet regarding road-building as a purely technical task, as can be seen later on. One other resolution of the Third Congress was that new roads should be built according to the principles of town planning.

The beginning of a dichotomy between town planning and engineering was expressed by Adshead himself who very clearly argued in favour of town-planning principles when he wrote that major roads should not go through residential areas.

Indeed the main traffic route, unless it is bordered with a parkway, is as destructive to the interests of a residential district as is a railway line (Adshead 1915: 167).

Motorization in Britain: the main issues

By far the most important issue was road safety. From 1923 onwards there was a growing number of parliamentary questions on this problem. The discussion continued with varying vehemence during the 1930s and 1940s. The number of road deaths had increased from 1,700 in 1919 to 3,000 in 1924; and had reached 7,300 ten years later, a figure which was only exceeded in 1965, apart from the years during World War II. Connected with that issue was the demand to improve existing roads and to construct new ones because many saw insufficient and poor-quality roads as the main culprit for the high number of accidents. Adshead argued, as many motorists and the road lobby did, that the existing speed limit was too low and that high speed by itself did not imply a higher risk of accidents. He saw the main danger in both bad road planning and road construction, including wrong design of curves and the lack of a standardized system of road warnings. However better roads reduced only marginally the number of accidents, at least according to the Ministry of Transport in 1937. They attributed 1.2 per cent of accidents to bad roads, whereas not surprisingly the AA attributed 12 per cent to bad roads.

The discussion on road safety touched another important topic namely the need to separate different transport modes. Alker Tripp, the Assistant Commissioner of the Metropolitan Police expressed his belief in 1938, which had been shared by many for several decades, that motor traffic can never mix safely with pedestrians and pedal cycles, and a civilized degree of safety can only be achieved by restricting this freedom of traffic movement.

Tripp's idea will be discussed in detail in Chapter 8. The separation of transport modes was a subject which became an issue after World War II and during the 1960s and was comprehensively outlined in *Traffic in Towns* by the Ministry of Transport in 1963. From the start the concept of separating different transport modes was seen in Britain as a policy to improve road safety. In Germany the reasons for the same policy were somewhat different as can be seen in Chapters 4 and 5.

Strangely enough the road lobby in Britain was against the idea of totally separated roads for motor vehicles which had already been suggested in 1903. According to Hamer the RAC and AA rejected the proposal of motorways because they feared that cars would be restricted to these roads (Hamer 1987: 37).

Another never-ending discussion concerned the width of roads. The road system in London was particularly bad. Some were in favour of wide roads because this led to the building of better-class housing. A distinguished official was of the opinion that roads had to have a width between 126ft (37.8m) between fences and 190ft (57m) between buildings (Brodie 1914: 269). About 50ft (15m) in the centre of these roads should be lined with trees and used as a public walk. One side of it would be bordered by a double line of tramways, and a stone-paved street, the other by a tar-macadam carriageway together with a normal footway. The traffic volume should be equally spread between the two types of streets.

Apart from the more technical discussions there were the publications by 'planners' which consisted of a mixture of engineering and planning ideas on this subject. There were contributions by Barry Parker and Raymond Unwin which will be discussed in the next chapter. In 1911, the American Charles Mulford Robinson, who had started his career as a reporter and editorial writer, published a book called *The Width and Arrangement of Streets*. The book was republished five years later with the new title *City Planning with Special Reference to the Planning of Streets and Lots*. In comparison with most of his contemporaries, Robinson had a very comprehensive knowledge of road and urban planning in the United States and Europe, especially in Britain and Germany. The reason was simple, Robinson had been invited by an American journal (*Harper's* magazine) to go abroad and study municipal improvements in Europe after completing a series of articles on the same subject in the United States. In his opinion good street planning was 'the product of philosophy, of sociology, and of economics as much as it is of engineering' (Robinson 1916: 89).

The sociological aspect was a completely new approach for the British and Germans alike. However, his ideas of street planning were adaptations of German design and garden city considerations which will be discussed in Chapter 4. In his opinion, residential streets should be planned in such a way that through traffic on such roads was discouraged and that the bulk of heavy traffic would be on a relatively small number of selected thoroughfares. His main argument was that often the major roads and boulevards were too narrow and the minor roads too broad. He wanted to devote 25–40 per cent of urban space to streets. He was against the artificial classification of roads but in favour of road classification according to their social and economic functions. Traffic was related to these functions, e.g. the retail business is associated with traffic streets. Street width should be related to the 'need' of streets. 'Some streets have community value, some have only restricted local value' (ibid., p. 71). Not only the number of houses per acre should be limited but also the amount of traffic for such streets. He was in favour of separate pedestrian footpaths between the gardens not only because it shortened distances but also because they were a delightful feature which would underline the 'restful rural charm' of neighbourhoods (ibid., p. 221).

However Robinson had no great impact on the discussion about the impacts of motorization in Britain. His approach was years ahead of its time and probably only partly understood when published.

The golden years of the motorist

The main battle during the late 1920s and '30s was between the road lobby, various pressure groups and the Ministry of Transport. The main topics were the Road Fund, petrol tax, compulsory insurance, driving tests and speed limits. The latter were highly controversial – ever since the 1903 Motor Car Act the argument had never stopped. It took years before a new Bill went through Parliament.

Lord Cecil of Chelwood had published his own Road Vehicle Regulation Bill in 1928. It contained four main suggestions:

1   a driving licence for every motorist;
2   the existing 20mph speed limit to be replaced by speed limits varying according to the type of vehicle, and vehicles were to be built so as to be unable to exceed the speed limits;
3   local authorities could install speed humps or similar devices in the road surface in order to force drivers to slow down (Plowden 1971: 141).

The representatives of the road lobby were horrified, especially by the idea of mechanical limitations of speeds which was in fact the first suggested traffic calming policy using physical measures in Britain. Although this Bill received some favourable comments it was defeated after much debate.

All this controversy culminated in the Road Traffic Act of 1930 which might be described as a success for the road lobby. The speed limit was abolished and the only achievement of the opposition was the introduction of compulsory third-party insurance. Plowden called the 1930s the golden age of the private motor car and the motor bus. The demand for cars had risen because of constantly falling motoring costs and prices. New cars were priced at half in 1930 of what they had cost in 1920; and the price had dropped further by about one-third in 1938. The average price of a car was £684 in 1920, £279 in 1930 and £210 in 1938 though wages were stable between 1925–1935.

Despite the long controversy before 1930, the new Road Traffic Act changed little in the main argument, which was that motor cars and motor cycles were responsible for 60 per cent of all fatal accidents. Road casualties continued to increase though the number of people killed

Table 3.1 Number of motor vehicles and road casualties in Germany: 1907–39

Year	Number		People	
	Cars	Motor vehicles	Killed	Killed and injured
1907	10 115	27 026	145	2 564
1909	18 547	41 727	194	3 139
1911	32 894	57 805	343	4 605
1913	49 760	77 789	504	6 817
1914	60 876	93 072		
	No data available from 1915 to 1920			
1922	80 937	165 729	No unified accident data from 1914 to 1936	
1924	130 346	293 032		
1926	201 401	571 893	2 398*	–
1928	342 784	933 312	4 589*	
1930	489 270	1 419 870	n.d.	
1932	486 001	1 653 297	n.d.	
1934	661 773	1 887 632	n.d.	
1936	945 085	2 474 591	8 388	182 214
1938	1 271 983	3 241 852	7 354	187 689
1939	1 426 743	3 705 111	no data	

* without Bavaria

Sources: Brandenburg 1936: 46; Statistisches Jahrbuch für das Deutsche Reich 1941/42.

between 1930 and 1932 had actually dropped. The most powerful pressure group, the Pedestrian Association, which had been formed in 1929 (Hillman and Whalley 1979: 17), mounted heavy propaganda campaigns on this subject.

As a result of public pressure a new Road Traffic Act came into force only four years later. The 1934 Road Traffic Act reintroduced the 30mph speed limit. The speed limit was to expire in 1939 but never did. The Act made driving tests compulsory and introduced pedestrian crossings.

Parallel to Britain, in Germany the street regulations of 1934 (Reichsverkehrsstraßenordnung, 28 May 1934) more or less abolished speed limits; only on some streets a speed limit of 40kph was laid down (Horadam 1983: 21). The relaxation of the speed limit continued until 1939. The speed limits agreed in the 1939 regulations were 60kph in built-up areas and 100kph outside, but these were reduced five months later to 40kph and 80kph respectively. Lorries and buses had to drive at lower speeds, in 1923 at 25kph or 30kph according to their weight, and from 1939 onwards at 40kph in built-up areas.

Tables 3.1 and 3.2 show the far higher degree of motorization in Britain. The German statistics imply that the number of people killed per

Table 3.2 Number of motor vehicles and road casualties in Britain: 1904–46

Year	Number		People	
	Cars	Motor vehicles	Killed	Killed and injured
1904	8 465			
1905	15 895			
1907	32 451			
1909	48 109			
1911	72 106	192 877		
1913	105 734	305 662		
1915	139 245	406 821		
1918	77 707	229 428		
1920	186 801	650 148		
1922	314 769	952 432		
1925	579 901	1 509 627		
1928	884 645	2 038 594	6 138	170 976
1930	1 056 214	2 273 661	7 305	185 200
1932	1 127 681	2 227 099	6 667	213 117
1934	1 398 425	2 405 392	7 343	238 946
1936	1 642 850	2 758 346	6 561	234 374
1938	1 944 394	3 084 896	6 648	233 359
1940	1 423 200	2 235 000	8 609	–
1942	857 700	1 840 400	6 926	147 544
1944	755 400	1 592 600	6 416	130 874
1946	1 769 952	3 106 810	5 062	162 546

Source: Plowden 1971: 456–7.

100,000 motor vehicles was higher than in Britain. The relationship was in 1936, the first year of real comparison, 238 in Britain but 339 in Germany. The reason for the higher accident rate in Germany may be related to the far higher percentage of motor cycles (46–48 per cent) than in Britain in which it was only 13 per cent of all motor vehicles in 1939. Even today motor cycles have the highest accident rate of all transport modes.

Conclusion

In Germany's large cities major street networks were planned and built from the second half of the eighteenth century which were modelled on Paris and Vienna. This very technocratic approach to street planning became more sophisticated with the publications of Reinhard Baumeister and Hermann-Joseph Stübben. Both already included street classification according to the traffic volumes. It was far ahead of corresponding English ideas of street planning. However this approach was soon challenged by Camillo Sitte who wanted to preserve the historic cities and was therefore not in favour of adapting the cities to the modern transport needs. In contrast to Baumeister and Stübben, Sitte's streets were curved, narrow, contained closed vistas and staggered junctions. Sitte's approach, which was adopted by many, dominated the discussions and implementations of new road constructions. Sitte's influence would continue far into the 1930s and had an impact on the reluctance to change the medieval city centres. His demands could only be successful because of a low level of car ownership and an excellent public transport system – the electric tram.

The development in Germany has to be seen in contrast to Britain where prior to the first quarter of the twentieth century the improvement of Britain's overall street network was hampered by confusing laws, regulations and responsibilities which made it difficult if not impossible to develop an efficient road network.

Bye-laws, which became common from 1875 onwards, for new urban street layouts were determined by health considerations. They took little account of the actual needs of traffic. They were not, as in Germany or France, a matter of overall street planning, engineering art or civic pride. It is important to note here that Paris had already undergone substantial changes in its urban structure when implementing the street plan designed by Haussmann between 1853 and 1869. Berlin's police department, which eagerly attempted to copy Paris, had given the go-ahead for a plan, which was similar in style to the French equivalent, including wide streets and boulevards (Hobrecht Plan: 1858–1860). The Berlin Zoning Law of 1875 allowed different street widths according to the land use of the town. Berlin's regulations were soon applied in most German towns, especially after Baumeister's publication on the need for master plans in cities in 1876.

The arrival of the car was treated in both countries with hostility. Yet, it seems the British Parliament made more fuss about it than the German one, which passed the first unified Act with reference to car use in 1909. The maximum speeds allowed in built-up areas were far lower than in Britain but the car ownership level in Germany was only about a third of that in Britain. The major changes in favour of the motor vehicle came with Adolf Hitler (see Chapter 7).

One of the most serious issues with reference to road transport was the increasing number of accidents in both countries. It was mainly this argument which made the fight by the road lobby for unrestricted speed limits such a long and intensive process in Britain. The period up to 1930 was characterized by a powerful campaign for improving road safety by stricter regulations, more sophisticated street layouts and suggestions which did not differ greatly from the German traffic-calming ideas of today. However the battle for road safety was lost with the 1930 Road Traffic Act. The road lobby gained its first major victory. Even so, the campaign to improve road safety did not stop and four years later a new speed limit in urban areas was introduced together with the compulsory driving licence.

Discussions on road safety were common too in Germany (for more detail see Chapter 5) but the battles appeared to have been not so fierce despite the far higher number of people killed per motor vehicle in comparison to Britain. This could be possibly explained by the economic and political instability which was foremost in people's minds.

In connection with the discussion over road safety, the argument concerning the separation of transport modes, which had already been suggested during the eighteenth century, strengthened in Britain. It was supported by both urban planners and engineers but had little impact on actual road designs. Despite passionate discussions about the need for more and better roads, relatively few new substantial road programmes were implemented, and certainly not on the scale suggested by some planners or as carried out in France, Germany and Austria. All major urban cities were lacking major arterial roads, notably London. It took until the 1920s before major urban road improvement and building were carried out.

4 The British and German garden city movements, their precursors and contemporary ideas

Barry Parker and Raymond Unwin developed a new approach towards roads

Plenty of books and articles have been written about the lives and work of Raymond Unwin and Barry Parker. In the context of this book, there appear to be several aspects which are important: the development of new street layouts, the functional division of roads, questions of road safety, the protection of residents from traffic, and whether Parker and Unwin were influenced by German speaking planners and architects to develop any of these approaches.

Parker and Unwin's major architectural and planning concept was derived from a critical and reformist 'Zeitgeist' about English society. They saw themselves as socialists who wanted primarily to improve living and housing conditions, though Unwin later had much wider objectives. Some of their ideas were the result of early and and intensive contacts with socialist and Marxist thought. This background helped to unfold their own ideas about society which were closely related to their philanthropic ideals. There had already been several successful practical, though isolated, attempts at creating better housing standards and new street layouts by the time Unwin joined Parker in his business as an architect in 1896 (Jackson 1985: 24).

The principal architect of Port Sunlight, William Owen, suggested superblocks as the main element of the model village, built in 1888. This design reduced the number of service streets. The superblocks were also pointed out by Berlepsch-Valendás, a member of the German Garden City Society. He was of the opinion that the design and grouping of housing in Port Sunlight with yardlike squares were the predecessors of the housing yards used later so successfully by Parker and Unwin in Hampstead Garden Suburb (Berlepsch-Valendás 1912: 103).

The larger housing blocks in Port Sunlight were filled with allotment gardens and therefore had a functional use (Creese 1966: 110). Many of the streets were curved, especially in the oldest part, the Dell. According to Choay, Port Sunlight had a circuit of main streets and a circuit of footpaths (Choay 1969: 30). It was the first housing development in Britain in which traffic was separated. The main question remains who or what was the source for this idea. The most obvious explanation may be that Owen copied the street layout of Paxton's Birkenhead Park near Liverpool which was designed in 1844, and was interestingly enough not far from Port Sunlight. Birkenhead Park also showed, apart from curving streets, an independent pedestrian footpath network. Paxton was a great believer in the separation of traffic flows, which can also be seen from his plan for the Great Victorian Way (see Chapter 2). There appears to have been a direct historic connection between the landscape school of gardening and an informal system of street planning.

Bournville in contrast to Port Sunlight was built (1879–95) much more with traditional bye-law streets, though it too had a few new features, such as the design of streets according to the contour of maps in preference to straight lining (Creese 1966: 111) or to plant a different specimen of tree along each street (ibid., p. 118).

Apart from Saltaire (1853–67), Bedford Park (1877–81), Bournville near Birmingham or Port Sunlight near Liverpool, there were many other less well-known examples of low-cost housing, such as Akroydon near Halifax and Woodland near Doncaster; though Bedford Park was built for the middle-income groups. In these suburbs or villages, many aspects were much further advanced than in Germany. To mention only a few, the more widespread implementation of constructing better and healthier housing for lower-income groups, which implied low-density housing, ample open spaces, provision of education and playgrounds. There was the common practice of locating the main rooms of the house towards the south whereas in Germany the main rooms still faced the streets independent of the location of the sun (Muthesius 1979: 68).

Parker and Unwin copied many elements of these early industrial villages (and from Bedford Park). They insisted on low housing density, designed roads according to the contours of the site, planted different types of trees or sometimes deliberately did not use trees so as to make a street look more interesting, attempted to separate or reduce traffic in the form of superblocks, different street widths, and footpaths.

Parker and Unwin saw, as did many of their contemporaries, the classical English village as a model for a community in a better world. It was the village green, the grouping of houses around it and the 'long wide village street bordered with clusters of cottages' (Parker and Unwin 1901: 91) which attracted them. It was a romantic picture full of harmony and beauty. The sentence 'as beautiful as an old English village' (ibid., p. 91) was the start of their first book and embodied their dreamlike projection of future settlement patterns. In addition, they saw in the Oxford and Cambridge colleges an image of communal life. But the high college walls were the main shortcoming of this concept. These walls would not allow any social contact with the outside world which was essential for both Parker and Unwin.

The repeatedly published themes of those early working years were the grouping of houses or clustering of cottages, of which some would open directly on to the roadsides and others would be situated around a village green. The importance of the location of houses with reference to the sunlight, the characteristic and the nature of sites, and the need for low density became objectives they never betrayed.

With reference to street layouts five aspects were important, such as an artistic design of roads and a relatively formal street layout. Sunlight was crucial, which would influence where and how roads would be located on a plot. Further, the cost of roads, which was determined by the number and the width of roads. But the objective was to remove the 'squalor of the back yards'.

In general roads and streets became integrated in the overall planning design of a housing settlement and lost their purely functional use. Parker and Unwin wanted to avoid streets which looked dreary and uninteresting. They intended to create additional open spaces. If cottages were built in blocks instead of a straight line the design of the street would look more interesting.

Parker and Unwin's understanding of the harmony of a village, their explanations about the laws behind it were oddly enough paralleled by the then still unknown Camillo Sitte in his book *City Planning according to the Artistic Principles*, 1889. The difference was merely that Parker and Unwin saw the ideal settlement to be a village whereas Sitte saw it to be a medieval or Renaissance city. These contrary starting points of the two nations appear to be one of the major differences between the continental and British approach to urban and road planning at that time. It is interesting to note that the first assignment with which Parker and Unwin made their reputation was to design a workmen's village near York.

New Earswick

In 1900 Parker and Unwin met Joseph and Seabohm Rowntree through the Garden City Association. The Rowntree brothers were inspired to build a village near York according to Garden 'village' principles. One year later they commissioned Parker and Unwin with the design of New Earswick. According to references, the plan of Bournville was most influential for this new commission (Jackson 1985: 49). However the first plan of New Earswick showed already significant differences from the Bournville design. It contained a mixture of curved and straight roads (see Fig. 4.1).

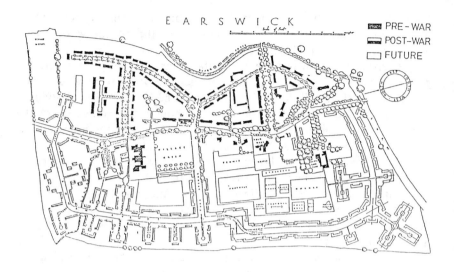

Figure 4.1 Plan of New Earswick, 1937.

The most striking difference was the use of narrow lanes which would connect the service roads with the houses. There is nothing left in the papers of the Rowntree Trust to elucidate the original meaning of these lanes. A possible explanation was that Parker and Unwin were trying to save costs by reducing the length of roads (Parker 1923: 8) and using narrow lanes instead. It may have also appealed to their sense of copying original village street layouts which consisted of lanes and different widths of streets. Some of the lanes appear to be the original country lanes which were integrated into the housing layout, but most of them were deliberately planned. The part of New Earswick which was first built showed an extensive use of footpaths; though in the later parts, built after World War II and in the 1920s, culs-de-sac were more common but footpaths still remained.

Today, New Earswick is still a remarkable village, and is probably the best preserved of all the existing housing developments built by Parker and Unwin. It has kept many of its original features. The footpaths are still in use and the service roads have been changed into pedestrian roads. As a result of pedestrianization a nearly perfect image of the time in which they were built has been preserved. The backs of the houses are served by access roads which also provide parking spaces for the residents (see Figs 4.2 and 4.3).

Figure 4.2 New Earswick: footpath access.

The garden cities: Letchworth and Welwyn

In 1898 Ebenezer Howard had published a small book called *To-morrow: A Peaceful Path to Real Reform*. It was republished in the same year as Parker and Unwin started to design New Earswick, under the new title *Garden Cities of To-morrow*. There is no need to discuss the background, the principles and the practical attempts of the garden city movement and its impact on British town planning, as they are widely known.

In 1904 Parker and Unwin won the competition to design the first garden city (Jackson 1985: 65–7). The first plan of Letchworth showed again different road widths, and the use of lanes, of which some were

Figure 4.3 New Earswick: residential street layout, pedestrianized since the 1970s.

previously existing lanes, such as the Icknield Way. The later plan of 1910 lost some of its clear street structure and abandoned many of the curving lanes (see Fig. 4.4).

Generally, the road system in Letchworth was relatively unsophisticated when compared with the system developed about fifteen years later in the second garden city built in Britain at Welwyn (see Fig. 4.5).

A comparison between the two road systems in Letchworth and Welwyn showed that in Letchworth, the difference in the carriageway width between the main roads and the residential roads was not very large. It is inexplicable that some residential roads had wider carriageways than the main roads (see Table 4.1). The width of the independent footpaths of which there were several in Letchworth is not included in Table 4.1. The early photographs of Letchworth show that many streets had wide side grass margins which today have largely disappeared. The relatively simple street network, and also the straight roads, today appear as one of the major weaknesses of Letchworth's road system. Motor traffic seems to be nearly everywhere and the only areas which have less traffic are Pixmore and Bird's Hill.

The street layout of Welwyn, which was designed by Louis de Soissons in 1920, shows the development in street designs (Filler 1986: 31). Interesting is the very wide main boulevard at the centre of Welwyn which appears to be a copy of continental boulevards. The division

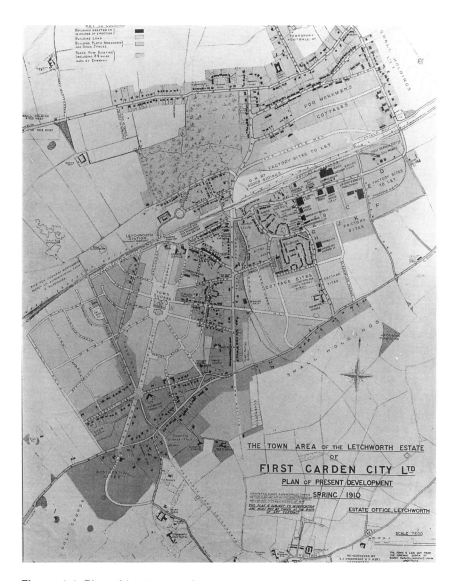

Figure 4.4 Plan of Letchworth Garden City, 1910.

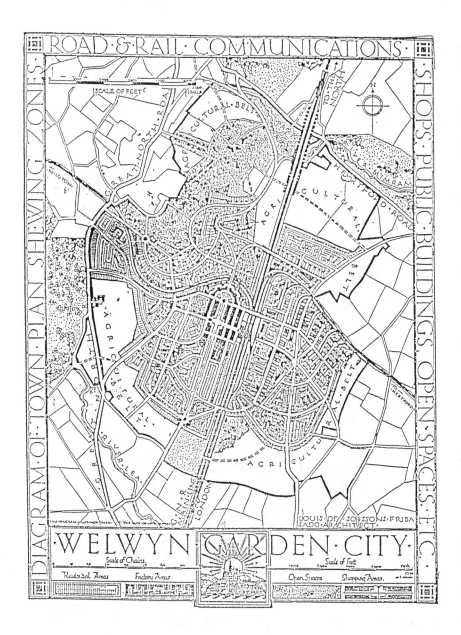

Figure 4.5 Plan of Welwyn Garden City, 1920.

Table 4.1 Variations in the street widths of Letchworth and Welwyn

Street type	Letchworth				Welwyn			
	Between boundaries		Carriageway		Between boundaries		Carriageway	
	ft	(m)	ft	(m)	ft	(m)	ft	(m)
Main street or promenade	100	(30)	27	(8.1)	200	(60)	2 × 18	(2 × 5.4)
Main highway	60	(18)	16	(4.8)	50–60	(15–18)	18–21	(5.4–6.3)
Residential road	50	(15)	24	(7.2)	40–50	(12–15)	15–18	(4.5–5.4)
Cul-de-sac	30*	(9)*					8–10	(2.4–3.0)

* Approximate.

Source: Purdom (1925).

between main and residential streets is clearer than in Letchworth. The narrow width of the culs-de-sac which even today are in many places kept at that standard is astonishing. They appear to correspond to a modern design form of traffic calming though parking is obviously a problem. However in most cases this could be solved with some careful consideration (see Figs 4.6 and 4.7).

After designing Letchworth, Parker and Unwin became increasingly interested in German planning and it has been argued that their later work was influenced by German ideas, especially by Camillo Sitte. In the next section these German connections and their impact on Parker and Unwin's planning will be examined.

The knowledge of German town planning strengthened new street designs for Britain

By far the strongest advocate of German ideas was Thomas Coglan Horsfall. Horsfall had tried since the 1880s to convince English local authorities about the methods German officials used to control their towns. His book *The Improvement of Dwellings and Surroundings of the People: The Example of Germany* made his name (Horsfall 1904: 9, 152). Horsfall had an astonishing knowledge about German planning, German local authority regulations and laws. He quoted extensively all the famous German language planning literature. For instance, Horsfall gave a detailed account of the laws of Saxony which stressed that streets should be as wide as needed for the requirements of local traffic (Horsfall 1904: 64). He also included a quotation by Stübben about the attitude of German planners towards streets in his chapter 'Narrower Streets for Towns in Germany':

Figure 4.6 Welwyn Garden City: typical small residential street, width about 2.50 m.

Broad streets certainly have advantages for traffic and in relation to supply of light and air; but, at the same time, they have disadvantages in respect of causing much dust and lacking shade . . . It was therefore seen that, while a well-considered net of broad streets for traffic must be provided for the extension plan, care must be taken to provide narrow *trafficless* [author's emphasis] streets, and thus to promote the building of the more desirable small dwelling house. Further, care must be taken, in preparing the plan, for an adequate supply of light and air in the interior of the blocks of buildings, for keeping streams pure, for open spaces and public shrubberies, and for the separation of manufacturing districts from the residential parts of a town (Stübben 1902: 28).

Stübben's attitude towards roads is important because Unwin used

Figure 4.7 Welwyn Garden City: footpath access to housing.

similar arguments later on.

From 1904 onwards contacts between the two countries intensified and we know about visits of the German Garden City Society to England in 1904, 1908 and 1909. One can also presume that from this period onwards, these visits were very regular, only interrupted by World War I. There was also a steady stream of British professionals and associations visiting Germany. There were several international conferences in 1906 (London), 1908 (Vienna) and 1910 (Berlin) in which the contacts between the two countries were strengthened.

Up to 1903 it appears that neither Parker nor Unwin knew anything about German planning or any other foreign experience. After that time, Unwin many have started to come into contact with Hermann-Joseph Stübben, who was already an honorary correspondent member of the RIBA by 1904, and information about Sitte's work and other important German planning ideas may have come from him.

In 1906, professional contacts between the two countries strengthened during the Seventh International Congress of Architects which was held in London. The conference was attended, apart from Unwin, by five Germans, some of whom played a crucial role in developing new thought and concepts in street layouts in Germany (H. Berlepsch-Valendás, T. Goecke, C. Rebhorst, J. Stübben and H. Muthesius). Stübben and Unwin participated in the same session and the affinity in thought between the two is remarkable. Unwin had by then already acquired

sufficient knowledge about the school of Camillo Sitte. Unwin and Stübben had the same opinions about not laying down rules over whether to
use straight or curved streets, the need for different street widths, the use
of tree planting, designing streets according to the nature of the site, and
the grouping of houses (Unwin 1906: vi and Stübben 1906: iv). However
Stübben's approach of separating traffic and recommending traffic free
areas – the main area of the market places should be free of vehicular
traffic (Stübben 1906: iv) – demonstrated a more radical approach on
this question by him. The astonishing similarity in thought leaves no
doubt in my mind that both had been in intensive contact for some time.

Though Unwin copied several ideas from Stübben he converted them
into a British context; for example the argument used by Stübben in 1902
to promote the building of the more desirable small dwellings on trafficless streets was used in a related form by Unwin later on. He wrote in
Town Planning in Practice

It may even be wise where the main road runs through a building estate to
arrange those houses so that most of them front on to subsidiary drives, only a
few having a frontage direct to the main road (Unwin 1971: 317).

He continued in writing that the character of the motor traffic was
anything but desirable for residents; the dust, the noise, the smell are all
objectionable features.

This argument was repeated in the article 'The City Development Plan'
written in 1910 in which he concluded that for the simpler dwellings, the
location on the main roads will become less desirable and that such
dwellings should be located on minor roads,

where they will be free from the dust and noise of traffic and where the amenities
of the site can more easily be preserved (Unwin 1910: 254).

This article is also most interesting with reference to his ideas of road
design and planning. Many foreign streets were in his opinion too wide
and there were also too many streets. He repeated his demand that
British roads had to have variations in the street widths in order to carry
out sensible town planning.

By 1910 Unwin was not specifically formal on how many different
types of roads were needed in a town. His opinion was that there should
be room for nearly every size of road, from the 20ft (6m) road to the
150ft (45m) multiple track road (Unwin 1910: 256). The main highway
would act as the framework for urban design. He pointed out the importance of different road categories which should form a relationship with
each other. He also referred to the need for radials, cross diagonals and
ring roads, again a very German characteristic.

He suggested a possible classification of streets into several categories
according to their function:

1 minor roads: width about 13ft (3.9m), and the width was to be increased to 24ft (7.2m) for turning places;
2 more important minor roads: width between 24 and 36ft (7.2–10.8m);
3 the main highways: width between 60 and 100ft (18–30m);
4 in a few cases even wider roads according to the amount of traffic were needed.

Some main highways should even be wider, allowing for different tracks, such as one for high-speed vehicles and one for local stopping traffic. For tramways he suggested wide grass margins.

An even more critical view on German street planning was expressed by Unwin in his articles of 1911 and 1914. He attacked the entirely informal arrangements of streets which some German towns were adopting and was of the opinion that such a conscious effort at accidental irregularity was unlikely to be the basis of any good planning design (Unwin 1911; 1914). This was a clear disapproval of the 'quasi-romantic school' led by Sitte and Goecke, a criticism shared by Stübben. Unwin was convinced that the formal street layout was better:

I believe that with care and imagination beautiful streets can be built on quite formal lines by simply manipulating the building line a little (Unwin 1911: 137).

In 1914 he also developed a new idea about the character of main highways. This thought came close to the concept of motorways which he knew from his visit to Chicago in 1911. He suggested that highways should as seldom as possible have crossroads because they would increase the danger of collisions and also delay the traffic (Unwin 1914: 32).

Between 1914 and 1923 Unwin did not significantly alter his attitude towards road planning. A major change came later in his life with much closer contacts to the United States – his daughter Peggy had married a publisher in Toronto (Jackson 1985: 162) – where he experienced car traffic at a volume not known at that time in Europe.

Another question is how German planning ideas influenced the practical work of Parker and Unwin. This will be discussed in the context of the design of Hampstead Garden Suburb which has been seen by many as the most 'German' layout for which Parker and Unwin were responsible.

Hampstead Garden Suburb

In 1906, Parker and Unwin got the contract to prepare a plan for a garden suburb in Hampstead (Jackson 1985: 84). The first detailed plan which came from Parker and Unwin's office had already been drawn in February 1905. It was the work of the architects Baldock and Buxton and not by Parker or Unwin (Creese 1966: 223) (see Fig. 4.8). In 1906

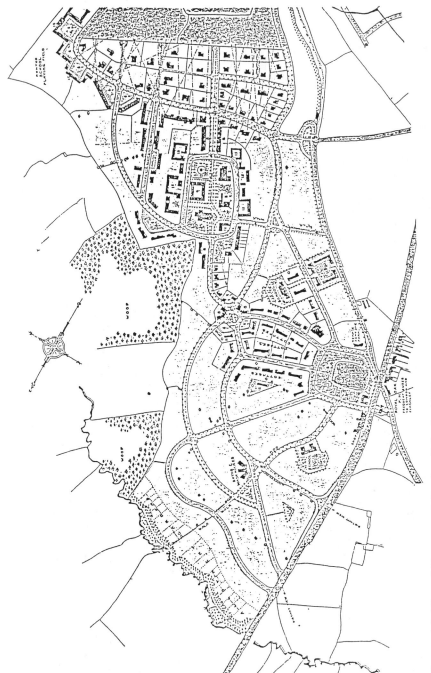

Figure 4.8 Plan of Hampstead Garden Suburb, 1905.

Figure 4.9 Plan of Hellerau Garden City, Dresden, 1906.

Edwin Lutyens was appointed as consulting architect and new plans were designed in the following years. The accepted plan was prepared in 1912 and was heavily influenced by Lutyens.

An Act of Parliament was obtained to change the bye-laws for the street width in Hampstead Garden Suburb. After the change the residential roads could be designed at a minimum width of 40ft (13.2m) and if wider roads were wanted grass margins and trees could be planted on each side to narrow the carriageway. With the change of the existing bye-laws, culs-de-sac were now allowed (streets not exceeding 500ft) at a width of 20ft (6m), with a carriageway of 13–14ft (3.9–4.2m) if houses were not closer than 50ft (ibid., pp. 308–9).

Creese mentioned that Unwin wanted to create a road system which benefited those living in a community instead of those passing through (ibid., p. 239). Unwin's desire to protect residents from through traffic brought a new dimension into British town planning and could be seen as a German influence. Yet Unwin did not quite succeed because Hampstead Garden Suburb had been plagued during the 1980s by a high volume of through traffic.

As we have seen in the previous section, Parker and Unwin were in favour of more formal, geometrical street layouts. Several Germans, such as Baumeister and Stübben, noticed this too and found it a great relief compared to the German practice. But the first plan of Hampstead Garden Suburb designed in 1905 was very informal and indeed very different from the plan of New Earswick and – even more so – from that of Letchworth. The majority of streets in Hampstead were curved (see Fig. 4.8) and at first glance there seemed to be some strong similarities between the plan of 1905 and German plans with strong Sitte-esque elements, for instance the plan of the German garden city Hellerau designed only one year later, in 1906 (see Fig. 4.9).

However, it is more than likely that Parker and Unwin may not have even read Sitte's book by February 1905 when the first plan of Hampstead Garden Suburb was completed and could therefore not have used Sitte's design elements for the first plan. This assumption can be confirmed by a closer examination, the plan of 1905 showed few if any typical Sitte-esque elements. In comparison to Hampstead, the plan of Hellerau, showed the typical staggered junctions which were used mainly in order to secure the closing of street vistas. Hellerau also had several irregular road junctions which were widened to form small squares. Both staggered junctions and irregular road junctions were not used in Hampstead. There was undoubtedly one element which can be mistaken as Sitte's influence and which was used in Hampstead Garden Suburb, namely the objective of producing 'vistas', groupings and street pictures. Yet these techniques had been used before in Letchworth; 'vistas' played a crucial role (see Fig. 4.8) but not so much 'closed vistas' which were a typically Sitte-esque element.

It is also possible that Parker and Unwin were not even very much convinced about the first plan, and that they may have welcomed the more formal layout developed by Edwin Lutyens in the coming years. Unwin made some disapproving remarks about English surveyors who tended to used curved instead of straight roads (Unwin 1971: 260). This apparent English trend was derived from landscape architecture which Unwin described in *Town Planning in Practice* (ibid., p. 27) and had little to do with German influence. Muthesius pointed out the similarity of Sitte's design features and the design layout in Bedford Park but the latter was built before Sitte had even published his famous book.

The informal street layout which was used not only for the first plan of Hampstead Garden Suburb but also for other housing developments in Britain may have for years been mistaken as Sitte-esque influence. It was much more likely to have been the result of Britain's own tradition of informal street design.

By way of conclusion, it is far less clear whether the street layout of Hampstead Garden Suburb was as much built and designed as a German model as is often assumed. However, there is no doubt that by the time Parker and Unwin supervised Hampstead Garden Suburb they knew about German planning and had met German planners. We will return to Unwin's ideas on street planning and motor traffic in a later section.

The housing question and new ideas of street design in residential areas in Germany

In contrast to Britain, there was little political priority to improving the social situation for the lower-income groups and the working class in the first two decades after the formation of the German Reich. Between 1878 and 1890, the Anti-Socialist Laws were passed which suppressed any political socialist movements. It was only after 1890 that socialist ideas could be put into practice in Germany.

The most important new ideologies concerning the urban environment were the Garden City movement and what I have named the 'Quasi-Romantics'. Both were politically close to the socialist parties. The main objective of the 'Quasi-Romantics' was to reform the existing urban housing structure, whereas the Garden City representatives were not interested in this aspect. Planners, like Baumeister and Stübben, were certainly in favour of improving the housing situation for the working class and the lower-income groups, but both were engineers and belonged more or less to the establishment. They were not directly involved in the layouts of 'social' housing. Baumeister contributed little to street designs in these newly developed housing areas. In terms of street planning both ideologies were strongly influenced by Camillo Sitte, Theodor Goecke, and Karl Henrici. Karl Henrici was one of the closest followers of Sitte. He had been professor in Aachen in Architecture and Urban Planning since 1875. In contrast to many other architects he used culs-de-sac, e.g. Plan of Knurow 1905 (Fehl and Rodriguez-Lores 1983: 465).

Improvement in the urban fabric: the 'Quasi-Romantics'

The view of the Quasi-Romantics, such as Theodor Goecke, Walter Curt Behrendt, Hermann Jansen and many more was essentially a watered-

Figure 4.10 Typical residential street in the Margarethenhöhe Garden Suburb, Essen, built between 1909–1920.

down mixture of garden city ideas; there was the demand for open green space, healthy living, playgrounds for children and many new and modern aspects of urban planning, such as land-use plans, buildings at different heights (*Staffelbauordnung*). All these were combined with some romantic ideas of the past. It was the medieval town which was the model the 'Quasi-Romantics' aspired to.

There were a group of intellectuals who developed a new understanding of street division. The existing division into traffic and residential streets as proposed by Baumeister and Stübben was seen by them as only a technical concept. They advocated a categorization which was ideologically derived from an idealized medieval street network. It was firstly expressed by Theodor Goecke in 1893. He argued against the formal division between traffic and residential streets which fulfilled in his opinion no purpose because a so-called residential street could easily be converted into a traffic street (Goecke 1915: 4). He modelled his residential streets on medieval lanes, which implied that the character of some residential streets would not allow vehicle traffic. Although the Quasi-Romantics agreed that streets were the skeleton of the urban built-up areas, new forms of housing were seen as the main aspect of a modern and social society. Basically the Quasi-Romantics tried to improve the traditional tenement housing block. Tenement blocks were grouped together into more or less rectangular shape. This type of housing was usual in Berlin, whose particular construction rule (Bauordnung: 1853) together with a number of other construction laws (Baufluchtliniengesetz: 1875 and Polizeiverordnung: 1897 and 1898) made different design forms difficult if not impossible, but it was common too in other German cities.

In Goecke's opinion the traditional housing block needed some reform. One of his main arguments was to design large inner courtyards in the housing blocks, in order to improve the cramped housing conditions. Low-density living was depended very strongly on how much open space was allowed for each block, what form the block had (closed or open), and how many storeys high they were built (Goecke 1918: 156).

Goecke's main idea was to develop urban housing blocks which consisted of different building heights. His principle was to relate the width of the street to the height of the building. The lower the height of the buildings the narrower the planned roads (Goecke 1915: 4). A traffic street would surround a four- or five-storey-high housing block, and inside the block, the buildings would become lower and the streets narrower. The symbolic similarity to the medieval town is striking, with the highest buildings representing the medieval walls. Though the idea of different heights inside housing blocks was not totally successful at the time, it was applied later in different forms in some of the public housing estates, 'the Siedlungen', built during the Weimar Republic.

Yet most of his other ideas were copied, and he was supported by many architects. The desire to develop new and different housing block forms led in the Weimar Republic to the large new public housing estates which were built in terraces (Reihen-Siedlungen).

There was another idea which became popular, which was possibly directly derived from England. Traditionally, the living quarters were always designed towards the major streets independent of the direction of the sunlight. By the turn of the century, several architects changed the design of the major living quarters and located them towards the 'quiet inside yards'.

New forms of housing block were developed everywhere and often new street designs were part of it. Often street regulations allowed the building of housing blocks which contained large yards. Normally they were filled in with high-density, low-quality housing. The Quasi-Romantics introduced changes in the traditional housing blocks. The number of housing blocks which were built around and just after 1900 with large inside open spaces including children's playgrounds – in traditional housing blocks children were not allowed to play in yards – and other facilities for residents is surprising. All these yards contained footpaths or residential streets, often connecting major streets.

These new possibilities of creating healthy high-density housing were aided by the success of housing associations such as the Berliner Bau- und Wohnungsgenossenschaft or the Beamten-Wohnungsverein. Some had been formed as early as 1847 (Hartmann 1976: 22).

The German Garden City Society

The German garden city movement was formed in 1902 when Heinrich Krebs, a business man, introduced the English garden city idea. It has never ever been clear whether Krebs only brought back Howard's book or whether he had in addition personal contacts with some of the representatives of the British garden city movement. Whatever the true story was, the English idea fell on fertile soil. The brothers Heinrich and Julius Hart, Bernhard Kampffmeyer and others had already been part of a romantic literature group in Berlin Friedrichshagen originally formed in 1888. At the turn of the century, some of this group decided to create a new idealistic community in Berlin Schlachtensee, called the New Community (neue Gemeinschaft), in which everybody was supposed to live in harmony with each other (Hartmann 1976: 28). This highly idealistic group had not been very successful in achieving their own objective; but a German garden city idea was something slightly more realistic and the New Community changed its name into the German Garden City Society (Deutsche Gartenstadtgesellschaft).

There were some other closely related reformist movements which should be mentioned here. The Garden Colonie in Oranienburg was formed in 1893 by 18 vegetarians from Berlin. Between 1894 and 1913 nearly 200 houses were built on the 55ha plot (Hartmann 1976: 35). There was also the industrialist Alfred Krupp who built better housing for his workers, modelled on English industrial socialist villages, between 1871 and 1874 (Klapheck 1930: 127). The more famous Krupp settlements – Alfredshof and Margarethenhöhe – were built between 1894 and 1920 (see Fig. 4.10). The designing of streets and squares was of crucial importance as part of the quality of the living conditions. The streets were mostly narrow, sometimes crooked, creating a cosy atmosphere, like a medieval town, and kept any form of through traffic out.

One of the main objectives of the Garden City Society was to create cheap and healthy housing. Clearly, a cheaper design standard for residential streets would lower the cost of housing. This kind of policy had already been successfully advocated and applied in England. The reductions in the number and in the width of streets, the designing of culs-de-sac, and footways also helped in reducing costs. The journal *Gartenstadt* was of the opinion that wide and expensive streets were the main culprit for high house prices (Kampffmeyer 1909: 68–70). However the promotion of new street layouts also included safety and environmental considerations. Kampffmeyer wrote that residential streets should be laid out in such a way that they could not be used by through traffic.

The residents will be very grateful if they do not experience the noise and dust of traffic (Kampffmeyer 1909: 68).

The normal width of such carriageways could be 4.50m or 5m and the pavements was calculated to be 1.50–2m.

The first German garden city, which was built as closely as possible in accordance with Howard's principles, was Hellerau near Dresden. All housing developments built by the Garden City Society were called 'garden cities' but all of them were garden suburbs, and some of them were very small indeed.

Karl Schmidt, who had a joinery workshop for handicraft art in Dresden, was the founder. He wanted to move his large workforce into the countryside and he provided the necessary funds (Hartmann 1976: 47–8). In 1906 Richard Riemenschmid took over the overall planning and construction started in 1909 (Kampffmeyer 1918: 336). Riemenschmid was an admirer of Camillo Sitte as can be seen from the overall street plan (see Fig. 4.9). Bernoulli, though himself a follower of the Garden City Society, made some outspoken remarks about the street layout of Hellerau.

In the department of urban planning in our seven technical universities, planners develop aesthetic guide lines which are all derived from the Middle Ages . . . and therefore we experience this strange play that many of our newly formed Garden Cities, including Hellerau, look as if they were built over the years into the past. Breaks, bellies, indented and jumping out corners are seen in the streets like in medieval villages and towns . . . (Bernoulli 1911: 111).

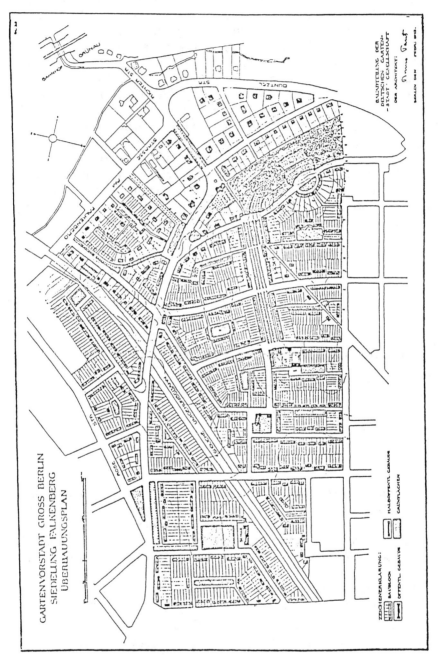

Figure 4.11 Plan of the Falkenberg Garden Suburb, Berlin, 1913.

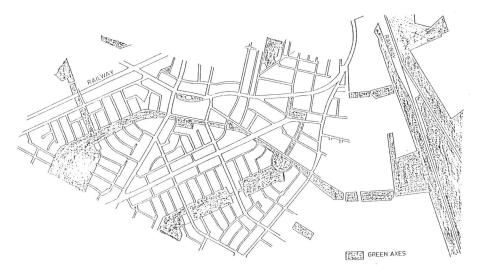

GREEN AXES

Figure 4.12 Plan of a garden suburb, Berlin, 1915.

The Garden Suburb Falkenberg designed by Bruno Taut and built in Berlin in 1913 had much straighter streets. Here streets were divided according to Goecke's ideas. Most of the residential streets surrounded the housing blocks. The centre of the blocks consisted of small squares or long rectangular squares (or a short green axis) which were surrounded again by even narrower streets which would connect to the residential road (see Fig. 4.11). Such street layouts clearly protected the residents from wheeled and later motor traffic. Walter Curt Behrendt's description is well suited to describe the street layout of Falkenberg:

> The traditional street pattern is replaced by a large network of traffic streets and open spaces. The mesh is filled in with more or less closed smaller systems of residential streets with play areas (Behrendt 1913: 101).

Related street layouts were used in many German garden suburbs (see Fig. 4.12). There is some similarity between the German street layout and the street layouts used in English garden cities and garden suburbs but the use of culs-de-sac was not so common in Germany. As in England narrow streets were mainly justified on economic grounds and this aspect was particularly important in Germany.

The Garden City Society had less than 10,000 members even at its height in 1915. The built and planned garden suburbs were very small; the largest one which was planned in Nuremberg was to have no more than 2,000 dwellings. Nevertheless it was intellectually a powerful organization. The ideas of the Garden City Society were carried on and so was its street design. It again became popular – though in a somewhat different form – both during the Weimar Republic and the Third Reich.

Conclusion

The planners and architects of the English industrial model villages developed the foundation of a new street layout. They started to design streets according to their own aesthetic values and, only as a second priority, according to their traffic function. These attempts influenced Barry Parker and Raymond Unwin and they were able to develop the street layout further in the English garden cities and garden suburbs.

There was great interest from some English architects and planners, especially Raymond Unwin, in the German planning approach which also included the important aspect of street planning. There was a strong affinity between Unwin's road concept and Stübben's who was one of the most distinguished German planners. Stübben was already in favour of separating traffic modes and he suggested that specific residential streets and city centre streets should be more or less freed from wheeled and motor traffic. Unwin undoubtedly adopted later some of Stübben's ideas and developed them into a British approach to road design which protected residents from the adverse effects of wheeled traffic. However, at least in the early years, pure design features to beautify roads were of crucial importance to Parker and Unwin, but there were also financial considerations which forced both to invent more sophisticated street layouts.

As in Britain, in Germany, new street designs were developed by supporters of the Garden City Society or the related movement, the Quasi-Romantics. The difference between the two countries was that, in terms of street layout, Germany was technically much further advanced and street widths according to land use and traffic flows were already common. Thus they could develop a whole range of new ideas, which were impossible for designers of the British garden cities or garden suburbs because they had to fight a much more fundamental battle against bye-laws.

In Germany the preferred narrow street widths of residential roads and the designing of footpaths, which were part of these layouts, were largely influenced by Sitte and Goecke. They were an attempt to recreate medieval street patterns. Apart from traffic considerations, the financial aspect was of crucial importance in Germany. Most of these newly planned residential areas included culs-de-sac and independent footpaths, though the use of culs-de-sac was not as common as in Britain. Other design features were green axes and/or squares, which would be interwoven within the residential areas (see Figs 4.11 and 4.12). They would either be part of the narrow residential roads, which were often only

Figure 4.13 Typical residential street layout in the 1930s, Brighton. The footpath is separated from the carriageway by a wide grass margin.

used as footpaths, or would have the function of wide pedestrian axes on which the school and the church would be located.

In both countries, the first two decades of the twentieth century were the most creative period in developing new street designs for residential areas. That is particularly valid for Britain where the street layouts designed by Parker and Unwin influenced most street designs in the coming decades, not only in the social housing estates but also for speculative housing (see Fig. 4.13). It is interesting to note that in future years new ideas in street design were experimented with first in the successors to the garden cities, the new towns. In Germany, the period of creative street planning ends with the 1930s, and early post-war Germany has, with a few exceptions, no new ideas in residential street layout.

5 The Weimar Republic: urban road transport policies and street design

Street design in residential areas

The Weimar Republic could be classified as an era in which new ground was broken by many new political and socio-cultural ideas which were brutally suppressed following the political victory by Adolf Hitler. Urban planning was even more vehemently seen as a force to create a new society than in the years before. Yet I believe that the Weimar Republic was to a large extent not so much a period of development of substantially new planning ideas, but more a time in which these could actually be put into practice. Many ideas on street layouts had been expressed before the 1920s. The influential planners had hardly changed. There was still the dominance of Baumeister, Stübben, Sitte and Henrici but also the ideas of the Garden City Society which influenced many young German planners and architects. Peter Koller mentioned that during the Weimar Republic the famous planning journal, *Der Städtebau*, lost more and more followers and closed down. This could indicate that the old guard became less influential.

During the Weimar Republic, the concept of garden suburbs was replaced by either independent satellite towns which were located about 20–30km from the main towns, or large housing estates on the outskirts of cities. It is interesting to note here the difference to Britain. Even in the past, the Germans had never designated independent garden cities. Hellerau was the closest model, but it was not really comparable with Letchworth or Welwyn because it was a very small settlement. The total space available for housing units in Hellerau was only 445, of which 345 had been built by the end of 1915 (Kampffmeyer 1918: 337).

The German Garden City Society was very successful in promoting the construction of small-scale garden suburbs, which implicitly kept the dependency between the suburb and the city centre. This dependency also existed with the satellite towns and the large housing estates built during the Weimar Republic, and the tradition was continued after World War II.

Figure 5.1 Römerstadt, Frankfurt, built after 1925. Typical residential street layout: footpaths run behind the houses and connect the housing estate with the green belt and the city centre.

Several of the avant-garde architects, such as Ernst May, Walter Gropius, Bruno and Max Taut, Martin Wagner, Hugo Häring, Hans Scharoun, Fred Forbat and Otto Bartning etc., had partly worked with garden suburb layouts and tried to develop them further, often in different directions. Martin Wagner and Bruno Taut had designed garden suburbs before and during World War I. Ernst May was strongly influenced by the English garden city movement and had worked together with Raymond Unwin.

A main objective of the architects of the Weimar Republic was to find new ways to design large social housing settlements. The major achievement was to overcome the traditional 'housing block'. The first attempts had already been made with the housing blocks built by the housing associations just after the turn of the century. Hermann Jansen in his land use plan for Greater Berlin included long housing blocks, two to three storeys high, with large inside gardens (superblocks) in 1909. This type of housing became legal as part of the official Berlin building regulations (Bauordnung) in 1925 and 1929 (Hofmann 1987: 400). A further development was the construction of housing in long and short terraces (Zeilenbau); terraces were regarded by many architects as a model for social housing.

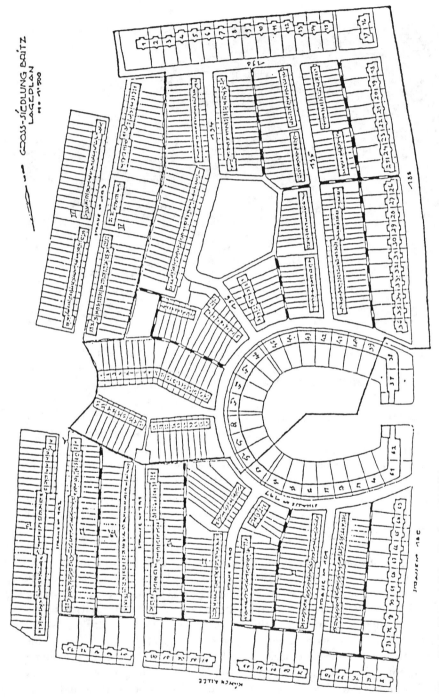

Figure 5.2 Plan of Britz, Berlin, 1925/26.

One of the finest new housing settlements planned and partly built during 1925–30 was the Römerstadt in Frankfurt (see Fig. 5.1), a settlement for about 50,000 inhabitants designed by Ernst May.

Berlin was another centre of this new type of social housing. There was the famous Hufeisensiedlung in Britz, built by Bruno Taut and Martin Wagner between 1925 and 1927; between 1930 and 1931 the settlement was enlarged, consisting then of nearly 2,000 housing units, of which 679 were single family homes (see Fig. 5.2). Onkel Toms Hütte was built by Bruno Taut, Hugo Häring and Rudolf Salvisberg between 1926 and 1932 and the Siemensstadt was planned by Hans Scharoun and other architects of 'the Ring' between 1929 and 1931 which had 1,800 dwellings. The Ring was a group of Berlin architects who were supporters of Das Neue Bauen. It was formed in 1926 and banned in 1933.

Social housing was strongly promoted in Hamburg too in the 1920s. Under the influence of Fritz Schuhmacher, who was responsible for urban planning, several smaller housing settlements were built, e.g. Dulsberg Siedlung 1919–31, Friedrich Ebert Hof 1928–9, Steenkamp Siedlung 1914–29, Jarrestadt 1927–30 and many more.

Yet, these settlements were not only concentrated in the large cities. Good examples can also be found in smaller towns, e.g. Celle, Georgsgarten built by Otto Haeseler, between 1924 and 1926, or Törten bei Dessau built by Gropius in 1926 (Pfankuch and Schneider 1977: 38).

Street layout was an important feature of these new housing estates. Yet the protective element for the residents which was present in the inside gardens of large housing blocks had to take other forms with terrace housing. In the latter, the footways would run behind the back gardens, mostly parallel to the residential streets. In most cases it was not a completely independent footpath network, and the footpaths would end in the residential streets. By and large, the principles of street classification developed and practised by the Quasi-Romantics and the Garden City Society stayed unchanged. The legal requirements of 1917 set out the function of streets in all newly built residential areas. Apart from the footways, called garden paths or economic paths (*Gartenwege* or *Wirtschaftswege*), there were residential lanes, residential streets and major streets which could be of different widths.

In some settlements, there was a clear separation between the street and the housing block or terrace. Access was given only by narrow footpaths, which were sometimes culs-de-sac. The housing blocks would face these 'pedestrian streets' which were often designed at right angles to the residential streets (Bauer 1935: 182). The settlement in Celle (Georgsgarten) built by Otto Haeseler had such a layout (Pfankuch and Schneider 1977: 38). Similar street layouts were used by Ernst May, Hermann Jansen, and several other not so well-known architects, e.g. Sorg in Nuremberg (Bauer 1935: 182). We can find them again in Britain

Figure 5.3 Siemensstadt, Berlin, 1929–31: footways inside the large yards, similar to Sunnyside, United States.

in some of the housing estates of the first generation of New Towns.

The ample use of green open spaces of the horseshoe form in Britz could be seen as a variation of the housing block, but its footpath network was not particularly sophisticated (see Fig. 5.2). In Siemensstadt, large housing blocks were part of the overall settlement, as suggested in the land-use plan of Greater Berlin by Hermann Jansen, and there, one could argue, residents were more protected from wheeled and motor traffic than in the narrow streets of the terrace housing estates (see Figs 5.3 and 5.4).

However most of the impressive settlements of the 1920s did not offer

Figure 5.4 Siemensstadt, Berlin, 1929–31: narrow residential streets. The residential streets are extremely narrow and parking on both sides is a problem.

any further development in protecting pedestrians from traffic from that already suggested when building the garden suburbs.

Let us look at one planner in more detail who was very influential during the Weimar Republic in terms of land use and street planning: Hermann Jansen.

Hermann Jansen and his road safety plans

Hermann Jansen, who was influenced by Sitte and even more so by Henrici, became famous from one day to another after he had won the first prize in the land use and master plan competition of Greater Berlin in 1909 (see Fig. 5.9). He had been an independent urban planning consultant since 1898 and became Professor of Architecture and Urban Planning in 1920 (Hofmann 1987: 389–94). According to Koller, Jansen was the only German architect who was primarily involved in land use planning and he employed about 8–9 people. He was familiar with the planning ideas of the United States and England (it is not clear whether he visited the United States), largely because he was editor of the journal

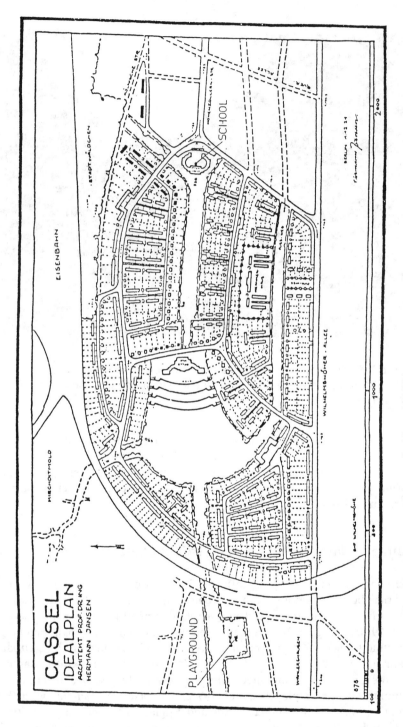

Figure 5.5 Jansen's plan of a residential area in Kassel, 1921.

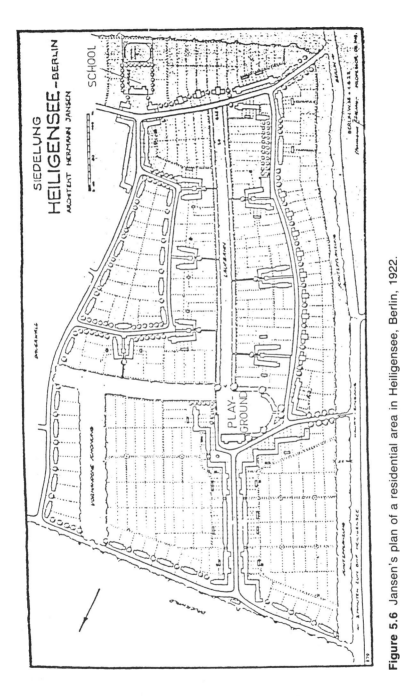

Figure 5.6 Jansen's plan of a residential area in Heiligensee, Berlin, 1922.

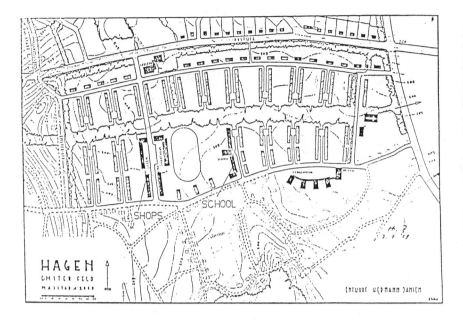

Figure 5.7 Jansen's plan of a residential area in Hagen, 1929: version 1.

Der Baumeister (since 1902) which stood in competition to the journal *Der Städtebau*. After his success in Berlin he became responsible for about seventy land-use and master plans in Germany and numerous plans abroad such as Ankara, Madrid, Montevideo, Bergen, Riga etc.

His main three principles of urban planning were expressed in his work *Die Großstadt der Neuzeit* (The Metropolis of Modern Times). These were: economics, transport and hygiene (Jansen 1917). He pointed out the danger of road traffic with reference to residential areas. Residential areas should be kept free of through traffic and sufficient underground lines should be built to keep the historic city centre free of it (it is not clear whether Jansen wanted to keep out all motor traffic from the city centre or only through traffic – the German is very difficult to understand). The influence of Sitte and Henrici is striking in his demand not to touch the old and historic (*Alte und Ehrwürdige*). He was displeased with parked cars which destroyed the attractiveness of historic squares. Generally, he saw cars as a health hazard because of the pollution and dust they created.

Very important for his land-use plans was the safety aspect. He attempted to eliminate 'black spots' in the historic street network and wanted to re-establish the rights of the pedestrians (Jansen 1917: 14–15).

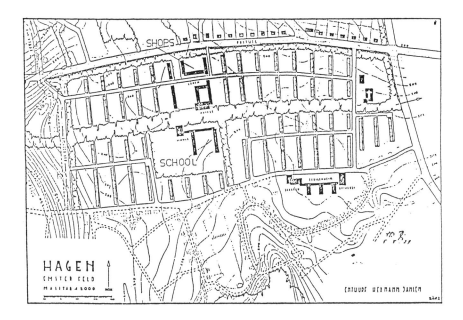

Figure 5.8 Jansen's plans of a residential area in Hagen, 1929: version 2.

The residential street layouts which were suggested by him were the closest copy of the Radburn layout the Germans ever developed. But again it has to be seen as a continuation of layouts which had been designed before the Weimar Republic. The garden suburbs planned by Taut and Wagner were clearly precursors of Jansen (see Figs 4.12 and 4.13).

Jansen's plans of Hagen showed the attempt to protect pedestrians and children from motor traffic (see Figs 5.7 and 5.8). It suggested the design of housing independently of the major traffic roads. He included a wide 'green' pedestrian corridor, which would connect the houses with all the main facilities, such as the school, shops and the church. Both plans achieved the characteristic that very few residential roads had to be crossed by going to school or carrying out shopping. The plan of Hagen was typical for Jansen and similar layouts can be found for many residential areas, e.g. in Nuremberg (Platnersberg), Kassel (see Fig. 5.5) or Heiligensee in Berlin (see Fig. 5.6). The idea of the 'green' pedestrian axes, first mentioned before and during World War I, continued in importance during the Weimar Republic and even more so during the Nazi period. Being a professor and employing young architects in his consultancy, he must have influenced many architects and planners.

Therefore it is not surprising to find similar planning designs later during the Third Reich.

It is not clear to what extent the Radburn design was widely known to German architects and seen as a model of modern street planning. By the end of the 1920s, several German planners had visited the United States and seen Radburn. One could also find articles about Radburn in the German journals.

By and large, there was little disagreement about the function of streets in residential areas and the right of independent footways in the new housing estates. There was also agreement over how the ideal street pattern of a newly built residential area should look. The residential streets stayed narrow and would not allow any through traffic to pester the residents. The main concern appears to have been the design of the housing in the form of relatively low-rise large blocks or terraces.

City-centre road traffic and regional planning

It seems that there was no unified approach in Germany on how to tackle the traffic problems in the densely built-up urban areas. Ideologically the conflict between the followers of Camillo Sitte, who were against street widenings, and the 'modern' urban and transport planners, who wanted to adapt the existing streets to the new demand for traffic, was difficult to overcome. Some of the most utopian ideas on city-centre traffic were developed in the late 1920s. It is interesting to note that much of the literature on that topic was written around that time. The main reason must have been the increase in motor traffic. Whereas in 1907 there were only 27,000 motor vehicles registered, by 1930 the number had increased to 1,420,000 motor vehicles (see Table 3.1, p. 43).

The increasing number of accidents made professionals and politicians also rethink whether the maximum speed of motor traffic was not too high. In 1923 the maximum speed had been increased from 15kph to 30kph in residential areas and could be increased to 40kph if desired by local authorities (Horadam 1983: 20). Hentrich was one of the planners who demanded a reduction of the maximum speed limit from 30kph to 20kph in residential areas. He believed that 30kph was too high to avoid accidents. He further argued that motor traffic should reduce its speed when it shared the street with other participants; this would also imply a reduction in noise and pollution.

As in Britain, there was no national street planning but the regional planning organization, Der Siedlungsverband Ruhrkohlenbezirk, formed in 1920, developed a unified approach to traffic in the Ruhr area. It was the second regional planning organization in Germany, modelled strongly on the regional planning organization of Greater Berlin formed in 1911.

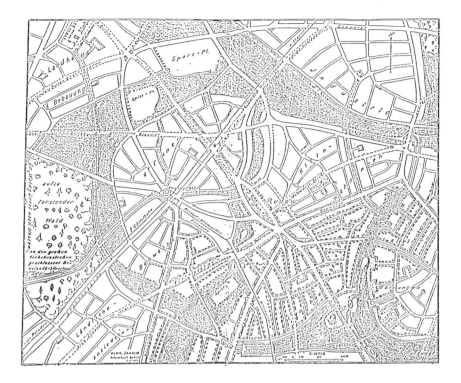

GREENBELT

Figure 5.9 Land-use plan of Greater Berlin: a settlement near Buckow-Rudow-Klein Glienicke, 1909.

Its main tasks were to set the alignments for through roads, for open green land, for housing settlements located on different local authority land, for transport corridors (*Verkehrsbänder*) and for airports. Transport corridors were kept free from any developments and could be used for traffic of all kinds. Local authorities had more or less lost their planning power on all these tasks. According to the regulations of the Siedlungsverband, not only were through roads included in the overall planning, but also other streets which were seen as important for the overall traffic in the Ruhr region. Figure 5.10 shows the overall planned network which consisted of roads going in east–west and north–south directions. One of the major objectives in street planning was to advocate bypasses which were to lead through the outer areas of towns. If this was not possible, ring roads should be built to relieve the city centres of the Ruhr cities from through traffic. Rappaport talked about the difficulties the Siedlungsverband had to convince local authorities

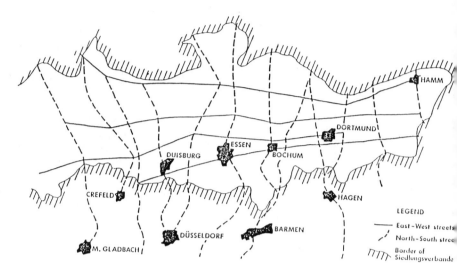

Figure 5.10 Main street network in the Rhine–Ruhr area, 1920s.

that major roads should not go through the town centres. The planned streets in urban areas had a width of 36m, of which 5m on each side were used for front gardens. All streets included tram tracks, which had to have as far as possible their own right of way.

As discussed in Chapter 3, new ring roads had been planned and built in other parts of Germany too. Some experts suggested that not only should the through traffic avoid the cities, but traffic generated inside the city should also be directed along the ring roads, e.g. Nuremberg.

Hermann Jansen was also in favour of relieving the city centre from through traffic. In his master plan of Nuremberg (1926) he demanded the diversion of the through traffic away from the historic city centre. The previous plan had included a straight north–south connection through the city centre with a tunnel under the castle. Jansen was of the opinion that such a route would redirect through traffic into the centre, both the construction of a major road and through traffic would destroy the historic city centre. Thanks to Jansen, the representatives of the city proudly pointed out to a group of English planners several years later their new traffic achievement.

Though there seems to have been little disagreement on the need for ring roads, the question of how to cope with the increasing traffic in the city centres was much more controversial. There were many planners who opted for street widenings. Some cities, Hamburg for instance, went uncompromisingly for new streets. Lichtwark formed the phrase 'Free and Demolition City Hamburg' (*Freie und Abrißstadt Hamburg*) in 1912

Figure 5.11 Arcades in Münster city centre. Open arcades were built to give more space to motor vehicles in the city centre.

(Klessmann 1981: 520). Not only the areas which had been destroyed by the big fire but also other parts of Hamburg were changed and demolished. Main roads were built everywhere in the city, mainly between 1867 and 1913.

Many cities were in favour of street widening because ring roads were, although regarded as a good solution, often too expensive and most medieval city centre streets could not cope with the increasing traffic. Some suggested that if the widening of streets was not possible then traffic should be separated into different traffic routes; others advocated taking away all the side walks and changing the ground floor into arcades and the first floor into shops. Some of the cities actually built such arcades though the shops stayed where they had always been (see Fig. 5.11). The plan of arcades was not uncommon and in some towns they were implemented as late as after World War II, e.g. Freiburg.

We find the first references to restricting vehicle traffic in the city centres decades before the Weimar Republic, but the 'need' for closing streets to traffic became stronger with the increase in motor traffic, which really only started to be of any significance during the late 1920s.

Baumeister referred to a street in Stuttgart which had been closed half-way down. The middle of the street was closed with a raised border, which was easy for pedestrians to walk over. Vehicle traffic had only access to the street from each end but could not drive through (Baumeister 1876: 94).

Police regulations existed for Essen's main shopping street in October 1904. According to paragraphs 1 and 27 of the police order, bicycles and cars were not allowed to enter the Kettwigerstraße between Burgstrasse and Bachstrasse; the Burgstrasse, the market and the Limbecker Straße were closed too. Servicing traffic was allowed. Disobeying the order was expensive, up to 60 German Reichsmark, not a small sum in those days (see Figs 5.12 and 5.13). However two years later restrictions were softened. In some parts of the Kettwiger and Limbecker Strasse traffic was now allowed but it had to drive at walking pace. In later years this police order may have changed again. Monheim quoted that the Limbecker Strasse in Essen was closed to wheeled traffic in 1929 (Monheim 1975: 129). The success of Limbecker Straße may have encouraged the city to close a further shopping street. According to Niehusener the closure of Limbecker Straße

appealed to shoppers and traders so much that the turnover in relation to street area of shops in that particular street was at the time among the highest in Europe (Niehusener 1974: 71).

Photographs exist from both the Limbecker and Kettwiger Straße, as late as 1938, showing the traffic signs which did not allow the presence of motor vehicles (Stadtplanungsamt Essen 1971: 15, 16).

Street closures of major shopping streets in the city centre of Cologne

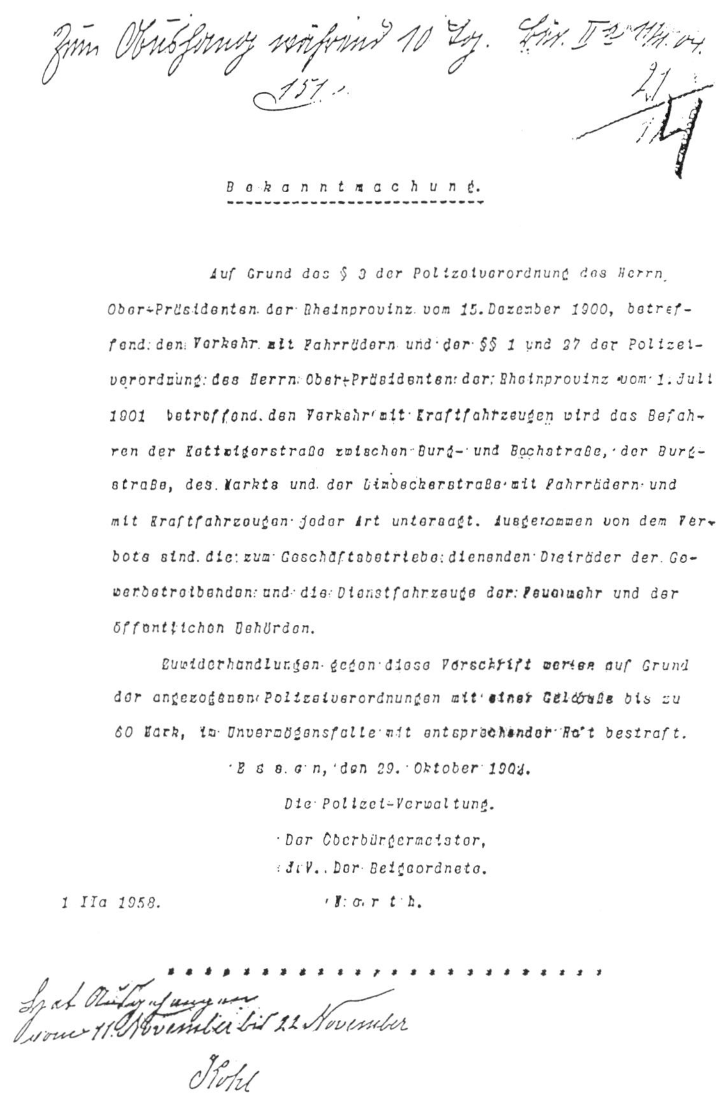

Zum Aushang während 10 Tg.

151

Bekanntmachung.

Auf Grund des § 3 der Polizeiverordnung des Herrn Ober-Präsidenten der Rheinprovinz vom 15. Dezember 1900, betreffend den Verkehr mit Fahrrädern und der §§ 1 und 37 der Polizeiverordnung des Herrn Ober-Präsidenten der Rheinprovinz vom 1. Juli 1901 betreffend den Verkehr mit Kraftfahrzeugen wird das Befahren der Kettwigerstraße zwischen Burg- und Bachstraße, der Burgstraße, des Markts und der Limbeckerstraße mit Fahrrädern und mit Kraftfahrzeugen jeder Art untersagt. Ausgenommen von dem Verbote sind die zum Geschäftsbetriebe dienenden Dreiräder der Gewerbetreibenden und die Dienstfahrzeuge der Feuerwehr und der öffentlichen Behörden.

Zuwiderhandlungen gegen diese Vorschrift werden auf Grund der angezogenen Polizeiverordnungen mit einer Geldbuße bis zu 60 Mark, im Unvermögensfalle mit entsprechender Haft bestraft.

Essen, den 29. Oktober 1904.

Die Polizei-Verwaltung.

Der Oberbürgermeister,

I.V. Der Beigeordnete.

I IIa 1958. Warth.

Hat ausgehangen vom 11. November bis 22 November

Kohl

Figure 5.12 Police orders in Essen: 25 January 1904.

Polizeiverordnung, betreffend Schrittfahren in der Kettwiger= und Limbeckerstraße.

Auf Grund der §§ 5 und 6 des Gesetzes über die Polizeiverwaltung vom 11. März 1850 (G.S.S. 265) und des § 144 des Landesverwaltungsgesetzes vom 30. Juli 1883 (G.S.S. 195) wird für den Stadtbezirk Essen folgende Polizeiverordnung erlassen:

§ 1.

In der Kettwigerstraße von der I. Dellbrügge bis zur Lindenallee und in der Limbeckerstraße vom Flachsmarkt bis zur Grabenstraße dürfen Kraftfahrzeuge, Fahrräder und Fuhrwerke aller Art nur im Schritt fahren.

§ 2.

Zuwiderhandlungen gegen diese Vorschrift werden mit einer Geldbuße von 1 bis 30 Mark, an deren Stelle im Unvermögensfalle eine Haftstrafe von ein bis drei Tagen tritt, bestraft.

§ 3.

Diese Verordnung tritt mit dem Tage der Veröffentlichung in Kraft.
Essen, den 5. Juli 1906.

Die Polizeiverwaltung.
Der Bürgermeister.
J. V. Der Beigeordnete: Werth.

Vorstehende Polizeiverordnung wird mit dem Bemerken veröffentlicht, daß die Bekanntmachung vom 29. Oktober 1904, betreffend das Verbot des Befahrens der Kettwigerstraße zwischen Burg= und Bachstraße, der Burgstraße, des Markts und der Limbeckerstraße mit Fahrrädern und mit Kraftfahrzeugen jeder Art außer Kraft tritt.

Essen, den 5. Juli 1906.

Die Polizeiverwaltung.
Der Bürgermeister.
J. V. Der Beigeordnete: Werth.

Figure 5.13 Police orders in Essen: 25 February 1906.

are also documented by the Historic Archives. According to a police order from 18 June 1921 the Hohe Straße and Breite Straße – two major shopping streets – were closed to all wheeled traffic including deliveries from 11.00 to 13.00 and from 17.00 to 19.00.

Motor traffic in the major shopping street (Sögestraße) in Bremen was only allowed to drive at walking pace up to 1930. The sign 'Drive at walking pace' – *Schritt fahren* – was removed in 1930.

Plans to pedestrianize the city centre or parts of it were suggested in Danzig in 1929:

Anyone who walks along the Langgasse in Danzig on a heavily trafficked day will of course immediately recognize the bad traffic conditions. In all other cities, especially in the Ruhr area, where the towns have not such high conservation value, it is urban planning policy that the traffic should be moved in case of difficulties into parallel streets. To do this in Old Danzig and at the same time to maintain the conservation of the historic town centre would cure the disease by killing the patient . . . One can not close the historic thoroughfare Langgasse–Langermarkt to vehicle traffic and at the same time scarcely 80m away tear down and widen the similarly historic Hunde-gasse and convert it to a modern shopping and residential quarter with an average height of 24m and heavy traffic. A conservation of old neighbourhoods is only radically carried out if vehicle traffic namely cars and trams are firmly excluded (Warnemünde 1929: 8).

After World War II, Danzig was rebuilt and the city centre pedestrianized.

Ernst May was of the opinion that traffic should be kept away from old historic urban districts. The effect of historic city-gates, which could act as traffic regulators because of their small entrances, was discussed.

In Jansen's plan for Hagen he suggested concentrating the business traffic on one side of the square, while the square itself should be kept totally for pedestrians.

Planning ideas for a new city centre

Bruno Taut's famous publication *Die Stadtkrone*, published in 1919, developed new ideas for the city centre. The city would be built at about 3km in radius (half an hour's walk) and was to be planned for about 300,000 people. The residential areas were to be built in garden city style. The centre point was the *Stadtkrone*, a building like a cathedral which was located at higher ground. Other representative buildings would be planned in close proximity to the *Stadtkrone*. The main traffic arteries were entering the city centre. But apart from a major road network, the residential streets were narrow, about 5–8m. His ideal city consisted of a park which radiated from the west, like a sector, leading outside from the city centre; it brought to it the forest, the fields and good air. It connected the heart of the city with the countryside. It was seen as the lifeline and included water basins, botanical gardens, flowerbeds, forests etc. In addition, the whole town was surrounded by a green belt (Taut 1919: 64).

Taut's idea of the green axes which would penetrate the city centre was used by Jansen in many of his land-use plans and can already be studied to some extent in the suburban land-use plan of Greater Berlin in 1909 (see Fig. 5.9). Reminders can be seen in Berlin Wilmersdorf (Volkspark). It was also used in his plan of Nuremberg. His general idea of the ideal city centre was to isolate the city centre from the outer districts by a green belt and also to surround the whole of the city by one. In addition, the city centre and outer urban districts would be connected by green axes of different widths. The inhabitants of the outlying parts and the city centre could walk and enjoy themselves in those belts undisturbed by traffic and noise. Like Taut, Jansen also designed major roads for motor traffic which would enter the city. He designed probably one of the finest urban motorways in Nuremberg, which would run along the abandoned and filled-in old Ludwig canal. In fact this road was indeed built in the 1980s.

Some of the planners and architects of the school 'Das Neue Bauen' developed more utopian ideas to solve the traffic congestion in the city centres, like the viaduct town suggested by von Doesberg in 1929. Others, like Scheibe, Hilberseimer and Hegemann, were in favour of upper-level walkways to create more space for motor traffic. Some of these ideas were not dissimilar to Corbusier's suggestions discussed in Chapter 2. The conflict between the followers of Camillo Sitte who wanted to preserve their historic city centres, and the avant-garde architects who wanted to adapt the city centre to modern technical achievements was much more evident in the city centres than in the newly designed residential areas. However, the 'modernists' made no significant impact on the city centres. In most cities no long-term perspective in terms of future traffic requirements was implemented, mainly because of their weak financial situation.

Conclusion

It was during the Weimar Republic that the first overall major road plans were developed and built in the Ruhr area. We also find the first motorway plan which will be discussed in detail in Chapter 7. Many planners agreed that through traffic should be kept out of the city centre.

The approach in the city centres was not consistent. It is quite likely that some narrow city centre streets were already closed to wheeled traffic from the second half of the nineteenth century. Therefore the idea of pedestrianization was not something new and was further applied during the Weimar Republic. We have evidence that streets stayed closed to wheeled and motor traffic during the Weimar Republic but we do not know if the number of streets closed increased. As in the past, the preservation of the historic city centre was demanded primarily by the followers of Sitte, who required the exclusion of through traffic and the retention of the traditional street network. This approach was questioned by some of the avant-garde architects.

New concepts for the city centre were developed by Bruno Taut and Hermann Jansen. But Taut's *Die Stadtkrone* was rather traditional and was not very much different from Jansen's idea. Ironically, Taut's *Stadtkrone* became the model for Nazi town centre and settlement centre planning. It appears as if the majority of the architects, even of the 'Neue Bauen', did not dare to change the historic city centre, apart from designs of high-rise buildings. We find few new ideas from them on how to cope with city-centre traffic in the existing city centres. It is therefore not surprising that little was changed and if so only in the form of street widenings. Some cities had a more radical approach than others. However the economic situation in 1923–24 and by the end of the 1920s proved also to be a major barrier in carrying out substantial street widening plans in the city centres.

The greatest achievement of the Weimar Republic was the construction of social housing. The traditional housing block of the nineteenth century was overcome by larger and low-rise housing blocks or terraced housing. Yet, the residential street layout was not developed significantly beyond that suggested by the Quasi-Romantics and the Garden City Society. In some cases the approach became more radical against motor traffic. However it is not clear whether this was caused by housing design itself being given the highest priority, or the most important issue being the protection of residents from motor traffic. Surely, in some instances it was both. The greatest merit of the Weimar Republic in terms of street layout was that the ideas which had been tried before often on a small scale in the garden suburbs were now generally accepted and used in all newly built housing estates.

The concept of green axes which would act as wide pedestrian footpaths and places of recreation strengthened and was implemented in many German plans. They were promoted by Taut's publication but they were also used by many other architects, including Hermann Jansen.

In particular, Hermann Jansen and his followers generated road designs which could be regarded as a further step towards improving road safety. He applied in a wider sense the Radburn principle. This is the more astonishing if one takes into consideration the extremely low level of motorization in Germany which was still below two million by the time Hitler came to power. Jansen's street layouts and also his superblocks were very similar to those developed somewhat later in the United States in Sunnyside and Radburn, which will be discussed in the next chapter.

6 Urban road transport in the United States and the influences on Europe: the importance of the Radburn street layout

Street planning in the United States

In the previous chapters we concentrated on the developments in Britain and Germany, but in many respects street planning in the United States was not so fundamentally different from the European examples. But the United States varied in one respect very significantly from Europe, namely in the number of motor vehicles which were produced and in use. It is therefore unsurprising that a 'new' street layout design to protect pedestrians from the flood of motor vehicles came from the New World.

Most urban and transport planners are familiar with the Radburn street layout. Its characteristic was to create both an independent network for motor traffic and for the pedestrian (or the cyclist); in doing so it reduced the conflict between these transport modes. Radburn was built in 1929. There are several interesting questions to be asked, for instance what was the background both in street design and in developing the Radburn idea in the United States? But let us first look at how streets were planned and designed in the United States.

Most North American towns were planned along gridiron street patterns which possibly had their origins in the Spanish colonial towns. This type of street layout was not changed for several centuries (Reps 1965: 12).

Probably in most towns the actual control over street and alignment was by no means clear. Reps described this situation for New York around 1800. In order to improve the situation, street commissioners had been appointed at least in the large cities along the East Coast at the beginning of the nineteenth century. However they were largely powerless if private owners subdivided land and designed streets according to their own liking. It took until the beginning of the twentieth century for official street plans to be established in most East Coast cities (Scott 1969: 4). By the turn of the twentieth century it was only Pennsylvania which had passed legislation that every municipality had to have an overall plan for its streets and alleys but Philadelphia had already had a tradition of regulated urban expansion and street planning, since its foundation in 1682. The gridiron street layout created street widths which were almost excessive by European standards. Also in contrast to many other gridiron street plans, which were later modelled on Philadelphia, the major streets were much wider than the rest of the street network (Reps 1965: 172).

The street plan of Washington, designed by L'Enfant, had by the end of the eighteenth century (1791) become a model for many other cities, such as Buffalo, Detroit, Indianapolis, Madison and many more. The planned avenues, which had plenty of space for walkways, including the main axis, were extremely wide and bordered with trees (160–400ft). Washington was modelled on Baroque street layouts, and as L'Enfant was of French origin, some of his ideas were influenced by Paris and Versailles, though he too had studied the street plans of many other European cities, including Karlsruhe, Frankfurt, Milan and Amsterdam.

Daniel H. Burnham had taken over the main responsibility for the construction of the Chicago Fair to celebrate the four hundredth anniversary of the discovery of the New World in 1894. He had fathered the City Beautiful Movement with this exhibition and it influenced many other cities in the coming years. Burnham and Edward H. Bennett also became famous for several overall street plans. Examples were the plan for Washington in 1901, which was largely a redevelopment plan of L'Enfant's original street concept, or the plan of San Francisco, designed in 1905, a year before the big earthquake (Reps 1965: 504–19). However the San Francisco plan was not to be influential in the reconstruction of the town. In contrast, the plan of Chicago, introduced to the public in 1909, was largely implemented. All three plans showed some similarities to the Haussmann plan of Paris. The gridiron plans consisted primarily of wide boulevards, diagonal roads, large squares and parks, including three major semi circular ring roads (Hines 1974: 328–9). It also proposed the first double-decker boulevard, with one level to be used for commercial and the other for regular traffic. By American standards a remarkable street hierarchy was proposed, at three levels. The through traffic was to be separated from the local residential traffic. The wide boulevards – parkways – the widest being 572ft, had been designed by Olmsted. The Chicago plan stimulated similar improvements in other cities, e.g. Minneapolis. It also had some impact on European street planning, especially in Germany. The Emperor of Germany was so enthusiastic about the Chicago plan that he appointed a commission to prepare a similar plan for Berlin (Ibid., 343).

However all these street layouts did little to protect residents or pedestrians from wheeled traffic and later from motor vehicles. Whereas the European medieval street layout would protect residents from such traffic to some extent because of its large number of alleys, lanes and narrow streets, the gridiron street layout had the opposite effect. Traffic could pour in and spread easily even into purely residential areas. Heckscher and Robinson talked about the 'deformation' of these plans by the automobile (Heckscher and Robinson 1977: 24). Yet, there were significant differences in the quality of gridiron streets. The street layout in New York was inadequate even by early nineteenth century standards because streets intersected too frequently and there was a lack of north–south arteries (Reps 1965: 299). The plan of Chicago was more sophisticated and included embryonic elements of traffic separation.

The American predecessor of the Radburn street layout was the romantic movement, which developed from the middle of the nineteenth century and had derived from English landscape gardening. The main characteristic in terms of street design was the heavily curved street which was in total contrast to the existing street plans. Such design was first used for cemeteries and later parks. The landscape gardener Frederick Law Olmsted and Calvert Vaux, an English architect, both became well-known for the design of Central Park in New York (1858). There they designed a street and path network which separated the different transport modes (see Fig. 2.3). They had an influence too on street planning in some suburban housing developments. One of the best-known suburban settlements designed by Olmsted was Riverside, Illinois near Chicago, in 1869 (see Fig. 2.4). He also planned Tomaco, the Western terminus of the Northern Pacific railway line in 1873. However this plan was quickly overturned by a more conventional gridiron street layout.

These plans showed for the first time some limited considerations for protection of residents from traffic. The plan of Riverside consisted of heavily curved streets. The street layout included some large green areas and there was a variety of street widths. Some plots were opened by much smaller streets. Culs-de-sac were used in Roland Park, Baltimore, which was planned by Olmsted in 1891. Hegemann also mentioned Roland Park and Black Rock in Bridgeport, which used the division of main roads and footpaths and the grouping of housing as part of their design (Hegemann 1925: 119–24). The plan of an industrial town for workers of the Apollo Iron and Steel Company, Vandergrift, north-east of Pittsburgh, Pennsylvania (1895) showed again Olmsted's familiar street layout including different street widths. Narrow roads gave access to inside housing blocks.

Similar street layouts can be seen for Oak Bluffs, Massachusetts by R.M. Copeland (1866) and Ridley Park, Pennsylvania (1875) also designed by landscape gardeners.

American landscape gardeners started to experiment with new ideas for street layouts. Most important here is to know that Henry Wright, one of the main planners of Radburn, was trained and had practised as a landscape architect and was certainly familiar with the suburban designs of the romantic movement (Churchill 1983: 208).

The impact of motorization and the issue of road safety: a comparison between the United States, Germany and Britain

Between 1908 and 1914, the mass production of cars in the United States changed both their price and their functions. Ford started to produce its first Model T in 1909 and this event changed road traffic very rapidly.

From 1915 onwards, the United States began the transformation from a rail-orientated country to a motorized one. Between 1920 and 1928, the railways lost 38 per cent of their passengers, whereas the 'Common Carriers' traffic had already lured away about one-third of the railway passengers in 1929.

By 1914 1.2 million cars were registered, and this increased to over 23 million by 1930. The acceleration in car ownership was dramatic, especially during the 1920s. Whereas in 1922, the annual sale of cars was 2,274,000, six years later the number sold had doubled to 4,455,000.

In contrast, Germany had only 50,000 cars whereas Britain had 106,000 in 1913 (see Tables 3.1 and 3.2). Even by 1930 the number had increased to just under half a million in Germany and to over one million in Britain. There were many reasons why the European countries, especially Germany, lagged so far behind. The main reason was clearly that cars in Europe were extremely expensive both to buy and run. This was firstly because they were individually hand-made and secondly because running costs, especially petrol costs, were high. There were also differences in the taxation of motor vehicles.

A strict comparison in terms of price and costs of cars between the three countries is difficult. In Britain, cars owned by the average member of the Automobile Club cost more than £300 just after 1900. Car prices were truly astronomical considering that only 4 per cent of the population in Britain left property worth more than £400 (Plowden 1971: 39). In 1912 when the Ford Model T came on the British market and was priced at £133, in comparison the cheapest car in Britain at that time cost £165 (Morris). But such cheap prices were an exception; most cars cost more than £200 (ibid., p. 102). One has to remember that the annual income of a bank clerk was £150 just before World War I. It was only after 1930 that British cars became cheaper in real terms.

In 1913 British firms produced 35,000 motor vehicles, the United States already produced in the same year half a million of which a quarter were produced by Ford. In Germany the car production was lowest, with about 8,000 in 1913 (Bardou, Chanaron *et al.* 1982: 15). Though British cars were made to high-quality standards, standardization in car production was much further advanced in the United States than in Britain or Germany. In 1938, six firms were responsible for 90 per cent of the car output in Britain of which two were American firms: Ford and General Motors (Maxcy and Silberston 1959: 15). Ford expanded to Britain in 1932 (Dagenham) and General Motors had taken over Vauxhall in 1928.

The German car industry was the least developed, largely because the period after World War I had been economically very unstable and was characterized by recession. Its output was very small and up to 1910 Benz was producing 50 per cent of all cars (Bardou, Chanaron *et al.* 1982: 15). The recession in 1923/24 stopped the production of four-fifths of all car manufacturers. In general, Britain was a much wealthier country and the average wages were higher during this period than in Germany; this may also account for the lower number of cars in Germany. Even the takeover of the German firm Adam Opel by General Motors in 1929 and the opening of Ford's branch factory in Cologne in 1925 (Yago 1984: 31) may not have been the turning point for car production in Germany, as Glenn Yago has argued. The real turning point came with the production of the Volkswagen after World War II.

The difference between the three countries was primarily a question of the price of cars, and running costs, in particular the price of petrol, and the average available income. Brandenburg pointed out that the Ford V8 was priced in America at $600 (1,500 RM) and in Germany at 5,000 RM. In addition, petrol prices per litre were four times higher in Germany than in the United States (Brandenburg 1936: 91).

Yet in terms of the quality of streets there were several parallels between Britain, Germany and the United States. Most urban streets in the United States were ill-suited for motor vehicles. Streets were narrow and badly – if at all – paved (Brownell 1980: 67). Germans reported from their journeys to the United States that the streets were partly in very bad condition. They frequently had large holes which were unthinkable in Germany, but nobody seemed to be concerned; even the pavements were often in an inadequate state.

Traffic of all kinds fought for space in the city centres. With increased motorization, the combination of streetcars and motor vehicles became a dangerous mixture. The dramatic increase in car ownership was hardly paralleled at all by new road construction in the more densely built-up areas of the cities along the East Coast. The choked city streets evoked discussions and plans about street widenings, extensions and motorways during the 1920s (Scott 1969: 187). There were the issues of absence of sufficient car parking, traffic management and traffic signals. Thus the motor car contributed significantly to the environmental and financial stress in urban areas. It was therefore not surprising to find that traffic restrictions or the closure of streets in the city centre of some North American cities appeared to have been quite common and were frequently mentioned in the literature from 1900 onwards; these measures continued during the 1920s.

Clarence Stein gave an eloquent insight into the traffic problems of the city centres in large American cities, which seemed to have been rather similar to their European counterparts.

Even in Los Angeles whose growth was coincident with the auto, the cars have multiplied faster than the streets have been widened. The end here is already in sight. There must come a time when every street in New York will be regulated as the streets in the financial district now are: individual vehicles will not be permitted to circulate through the business and industrial sections during the day (Stein 1925: 69).

Yet motor traffic had not only become a problem in many town centres; increasingly there was also the issue of road safety, which worried many Americans. Approximately 24,000 people were killed annually and 600,000 injured by motor vehicles in 1925 (Scott 1969: 187). As a result of these dramatic statistics a national conference on street and highway safety was held. Several initiatives were set up; for instance the Policyholders' Service Bureau of the Metropolitan Life Insurance Company undertook a traffic survey to demonstrate the best traffic control measures which could reduce deaths and injuries. It was largely the awareness of the danger of accidents which made some developers realize the necessity for a change in street layouts in order to protect residents from cars. The studies by Robert Whitten and Clarence Arthur Perry demanded that residential areas should be designed in such a way that they would protect residents from car accidents. It was this momentum which finally led to the construction of Radburn and later to the planning of the Greenbelt towns during the 1930s and '40s.

Regional planning and the Regional Planning Association of America

Urban and transport planning in the United States was totally different in character to planning in Germany or Britain. Planning had always played a much weaker role; even powerful and dynamic planners like Burnham and Bennett changed little of the image and function of planners. There was no strong housing reform movement, apart from some rudimentary slum clearances in a few major cities around the end of the nineteenth century, and even the City Beautiful movement did not develop into a strong social force (Sutcliffe 1981: 101–3). By the 1920s most city planners had become less reform-orientated and more committed to special and rather narrowly focussed skills . . . (Brownell 1980: 67).

If urban planners did not see themselves as a social force, transportation planners were even less likely to do so. American transport planners were not generalists, like most of their British and German colleagues, but specialists from very early on. They were involved in transport studies which had become common even before the 1920s. They included primarily data collection and surveys, but hardly any actual forecasts on the effects motorization could have on the urban structure.

Though the United States failed to rival Britain or Germany in urban planning, at the beginning of the 1920s the concept of regional planning gained momentum. It was largely caused by the desire to improve overcrowded and congested cities and the need for controlled urban growth, combined with ideas about constructing garden cities or garden suburbs. Regional planning agencies were created in several parts of the United States, of which the Committee on Regional Plan of New York and its Environs formed in 1921 was the earliest one. It was a voluntary advisory body and sponsored by the Russell Sage Foundation. It combined most of the important planners of the time (Scott 1969: 192–203). Several other regional planning agencies were set up. Some of them were county planning agencies, such as the Los Angeles County Regional Planning Commission or the Chicago Regional Planning Association.

Regional planning also included ideas and concepts about transport, but even by the late 1920s, hardly anybody had any idea of how the car would reshape the city. The Committee of the Regional Plan of New York and its Environs was led first by Charles Dyer Norton and after his death in 1923 by Thomas Adams, and Frederick Delano. The Scotsman, Thomas Adams – a friend of Unwin and Parker – had studied in detail the housing problems in Germany (Berlin) and Sweden (Robinson 1916: 55). He had also become well-known for his work in Canada.

Raymond Unwin went to New York to advise the Committee on the question of decentralization of housing and industry. He warned against the plan to increase road transport facilities, which was thought to cure congestion. He argued against the promotion of private cars. It was the argument about increased transport facilities which later exploded into a major controversy. The Committee decided in favour of a major promotion of road transport, and mass transit was to become of secondary importance (Kantor 1983: 189–92). It would be interesting to explore how far these New York planners were manipulated by people like Robert Moses who went for total promotion of car use and had little time for reformers. However, similar decisions were made in other cities, such as Los Angeles in 1924.

There was also another group of planners who were fiercely opposed to the Regional Plan Committee, the Regional Planning Association of America (RPAA), formed in 1923. It included planners, such as Lewis Mumford, Clarence Stein, Henry Wright, Frederick Lee Ackerman, Edith Elmer Wood, Catherine Bauer, Benton MacKaye, Alexander Bing, and a few more. A wide variety of professions could be found, and only about half of them were architects (Sussman 1976: 20). They were never a formal organization and communicated their ideas through books, articles, reports, and longed to put their visions into practice.

The RPAA believed in the garden city principles and we can assume that vivid dialogues between members of the RPAA and some British and German planners took place. Without doubt the ideas of many RPAA members were closest to the concepts of some of the most important British and German planners. According to the RPAA, garden cities were supposed to become the main settlement form for regional plans in America.

The RPAA shared the American admiration for European planning. Many members had studied British planning examples, especially the garden cities. Frederick Lee Ackerman had been sent to England in 1915 to study British war housing which was strongly influenced by the design of garden suburbs and garden cities (Lubove 1963: 39). Lewis Mumford had visited Raymond Unwin in 1920 and Unwin had close contacts with other members of the RPAA, such as Clarence Stein, Henry Wright, Catherine Bauer, Edith Elmer Wood and Frederic Lee Ackerman. In 1924, Stein and Wright visited England to study Letchworth, Hampstead Garden Suburb and the wartime munitions communities.

The RPAA represented a new form of garden city movement which had replaced the traditional one, formed by Ackerman in 1908 in the United States. Ackerman had been one of the very few foreign participants at the first Garden City Conference in Birmingham and Bournville in 1901.

The overall concept of the RPAA was complex.

Its philosophical origin went back both through Geddes to the French regionalist movement of the late nineteenth century, and to the early American concepts of balance and harmony between man and nature (Hall 1984: 47).

It was an idealistic group who 'sought to replace the existing centralized and profit-orientated metropolitan society with a decentralised and more socialized one made up of environmentally balanced regions' (Sussman 1976: 1). Truly a major task for a country such as the United States.

The RPAA was strongly anti-urban and saw the metropolis as a 'dinosaur', as Stein phrased it, which had failed in the most important planning issues of human society, such as housing, recreation, street system etc. (Stein 1958: 65). There was talk about the bankruptcy of the big city. It was their dream to create garden cities, and to move people out of these monstrous overcrowded cities. The ideal garden city should only have a maximum number of 50,000 inhabitants because more people would complicate and raise the costs of urban life (Schaffer 1982: 152). It is important to note the closeness of the RPAA ideology to Unwin's ideas at the same time. Unwin was obsessed by the idea of decentralization. The city represented for him the image of 'sprawling blight' and he wanted to re-establish the harmony between city and countryside.

Not surprisingly, the RPAA was critical of the Regional Plan Committee. Their criticism was directed not only against Thomas Adams, who they thought had betrayed his early vision of regional planning, but their major disagreement was against the proposed regional plan which was not directing growth away from New York but instead favouring further growth of the major city despite the planning of subcentres and satellite communities.

The members of the RPAA believed that a better future could be created with the support of new technologies, such as the telephone, electricity, and the car. Sussman was of the opinion that the RPAA members welcomed the car maybe too optimistically (Sussman 1976: 35). In that we can find some parallel with the beliefs of the avant-garde German architects and planners of the 1920s. Yet it is not at all clear whether the RPAA members welcomed it without any objections. In contrast to many other American planners, especially in contrast to the Regional Plan Committee, they foresaw the geographical effects cars could have on the urban development pattern. As it coincided with their vision of optimal regional planning, they obviously welcomed a flexible transport mode. However there was a much more differentiated view of car use in urban areas. Some of the members saw the adverse impacts the motor vehicle had in large cities which would further undermine the quality of urban life; they were aware of the danger to road safety in residential areas. It is also worth remembering that many members of the RPAA were influenced directly by Raymond Unwin who saw clearly the adverse effects of motor vehicle use in urban areas.

MacKaye had a very distinctive view about car use despite or because of his well-thought-out idea of interurban highways.

Motor traffic and pedestrian 'living' do not go together. To insulate each activity is a prime condition for speed and convenience on the one hand, and for safety and peace of mind, to say nothing of freedom from noise and carbon monoxide, on the other hand (MacKaye 1930: 93).

Possibly most of the RPAA members wanted to protect community life from the adverse effects of cars; MacKaye called it the adaptation of the motor car to an effective community life. The 'townless highways' and the housing estates built by the RPAA were the practical outcome of this desire.

The housing experiments of the RPAA: Sunnyside and Radburn and their influence in Germany and Britain

One of the first practical achievements of the RPAA was the formation of a private limited dividend company – the City Housing Corporation formed in 1924 – headed by one of RPAA's business men Alexander M. Bing. The main purpose of this organization was to create a garden city as Ebenezer Howard had planned for Britain (Stein 1958: 19).

The Corporation started to develop a derelict industrial site – called Sunnyside – for housing, close to Manhattan's business centre in New York. Its main objective was to provide low-cost but well-designed housing. The experiences gained in Sunnyside were to be used for the planned garden city.

Sunnyside consisted of one to three family housing blocks. The major disadvantage of the site was that the local government officials (Borough Engineers) had already planned a typical gridiron street layout. That implied that a major through road had to stay unchanged, despite the danger of increased accidents and the separation of the community into two parts (Wright 1935: 37). Though Wright had studied a garden community and could prove that housing developments based on a gridiron layout were needlessly expensive, they could not overcome the existing planned street regulations. These studies may also indicate some research on the existing literature written by Parker and Unwin, especially the famous publication by Unwin – *Nothing Gained by Overcrowding* (Unwin 1912) or articles published by Parker on the economics of street layouts based on culs-de-sac.

In Sunnyside, because of the rigid street layout, only a few culs-de-sac could be built (Stein 1958: 24). During the four years of constructing Sunnyside, different forms of housing blocks emerged. The characteristic Sunnyside blocks received, apart from private garden space, a communal green as the centre. This green included a network of footpaths which connected the different family units inside a housing block (see Fig. 6.2). Footpath connections between different blocks existed but one still had to cross the roads. By 1928 the Corporation had constructed houses for 1,202 families on about 77 acres.

As with the British and German Garden City Societies, the RPAA must have been rather effective in marketing their ideas and 'products'. Luckily enough, the completion of the first part of Sunnyside coincided with the International Town, City and Regional Planning and Garden Cities Congress in New York in 1925. Unwin, Parker and Howard were guests of the RPAA and we know the foreign participants were treated in style by them. We also know that Hermann-Joseph Stübben and Cornelius Gurlitt were present. Stübben remarked that Sunnyside consisted of

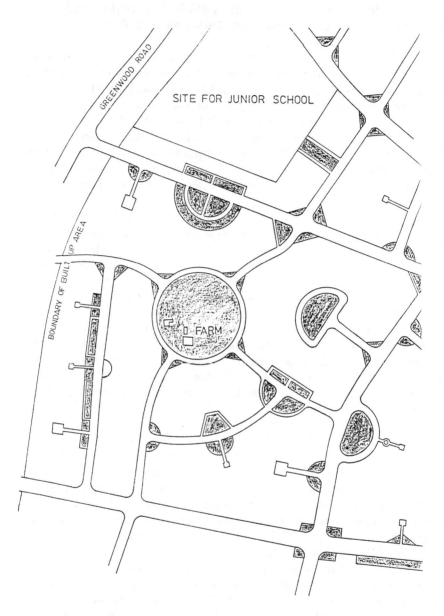

Figure 6.1 Parker's street layout in Wythenshawe, 1932.

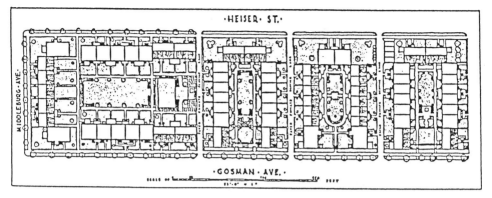

Figure 6.2 Plan of superblock in Sunnyside, 1927.

two and three storey high owner occupied housing blocks of more or less friendly appearance with large open areas inside them which were used for playing and other recreational purposes (Stübben 1925: 186–7).

Gurlitt pointed out that the President of the City Housing Corporation mentioned that Sunnyside was modelled on a German settlement (Gurlitt 1929: 27). There is no proof in the Anglo-Saxon literature I studied that this was actually the case, but the similarity with German housing blocks built after 1900 is striking.

After the financial success of Sunnyside, the City Housing Corporation wanted to fulfil their original plans, to build a garden city. The site chosen was only 16 miles from New York in Fairlawn, New Jersey, called Radburn. However, they soon had to give up the idea of a green belt and the attempt to attract any industries at the site (Stein 1958: 39). It was planned to build a new town which would consist of three neighbourhoods for about 25,000–30,000 people in total. Each neighbourhood had its own school and shopping centre. Since the start of the design and construction of Sunnyside, the members of the Corporation had developed their ideas further. Stein expressed the new emphasis:

We did not fully recognize that our main interest after our Sunnyside experience had been transferred to a more pressing need, that of a town in which people could live peacefully with the automobile or – rather – in spite of it. The limitation we found in the gridiron street pattern at Sunnyside, as a setting for safe motor-age living made clear to all the staff what we planners had long seen and had planned to eliminate (Stein 1958: 37).

Radburn contained several quite innovative elements. Let us firstly consider the sociological concepts of neighbourhood units. Related ideas had been discussed in the United States since the turn of the century (Robinson 1916). The main advocate was the sociologist Clarence Perry, also a member of the RPAA, who had worked as a social planner in

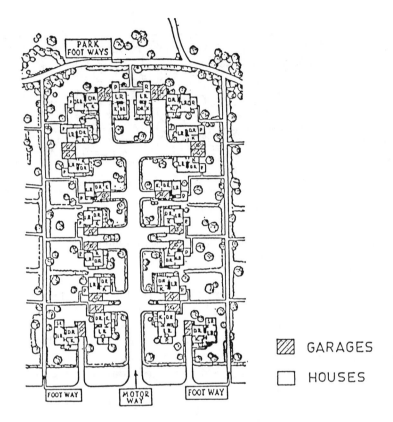

GARAGES

HOUSES

Figure 6.3 Typical 'lane' in Radburn.

New York and had been involved in the Plan for New York and its Environs. His opinion was that small-scale community facilities would create 'ideal' neighbourhoods. He suggested that cities should contain housing areas which were grouped around different cultural and commercial sub-centres, consisting of a school, community buildings, shopping centres, etc. Perry had lived in Forest Hills Gardens, New York, which was one of the earliest garden suburbs, designed as a neighbourhood in 1911. The original plans go back to F.L. Olmsted (Reps 1965: 525). The street layout was nothing particularly outstanding, but it gained reputation because of Perry's work.

Apart from the American neighbourhood concept, Radburn contained a street layout which was not common in the United States (see Figs 6.3 and 6.4). The street hierarchy in Radburn consisted of:

- service lanes for direct access to buildings;
- secondary collector roads around the housing blocks;

Figure 6.4 Plan of Radburn, early street design, 1925.

- main through roads, connecting the various neighbourhoods, districts etc.;
- express highways or parkways to provide inter-urban transport links.

Independently of the street network, a separate network of over eight miles of pedestrian and bicycle paths was designed. Over- or underpasses were constructed when pedestrian paths had to cross roads. All housing blocks had direct access by car. They were grouped together round large greens and parkland.

By American standards this road division for a relative large residential housing development was most unusual. The general practice was to construct all urban streets to serve as through traffic arteries (Sussman 1976: 25). The highway system was not sophisticated enough to do more than distinguish park and expressways from city streets.

As Stein pointed out, none of their ideas was specifically original. The separation of different transport modes in the form of independent roads or paths with over- or underpasses could at the time already be seen in the Central Park of New York which was planned and developed by F.L. Olmsted and C. Vaux in 1858 (Stein 1958: 44). According to Mumford, Olmsted did not

grasp the general significance of this separation for modern planning . . . but one can hardly doubt that Stein's daily walks through Central Park during his formative period encouraged him to hold to it tenaciously, once Radburn was built (ibid., p. 16).

There is no proof of whether the idea of functional division of roads was a copy of the Central Park street layout, or borrowed from Europe (which in fact was true for the layout of Central Park as well – see Chapter 2). In Germany, functional division of streets was common at that time, and it had been well established in the garden cities and garden suburbs in Britain. Several authors and Stein himself suggested that Parker and Unwin's street layout was in a sense the parent of Sunnyside and Radburn. Wright wrote that the culs-de-sac used in Radburn were derived both from the current English practice and their own experiment in Sunnyside (Wright 1935: 42).

There is evidence that a detailed design for Radburn was already available in 1925 and was shown to a wide audience of British and German planners during the International Town, City Regional and Garden Cities Congress. Gurlitt wrote about Radburn that it was built like a European village with the design standards which had been common in Germany before World War I. He continued that by German standards it was far too old-fashioned. It would not even be admitted at any international or national planning competition (Gurlitt 1929: 30). He was wondering why in a country as technically advanced as the United States, Radburn was seen as an expression of modern times and as an example of urban culture. Though Gurlitt was not impressed by Radburn's housing design – and neither was Stübben – he was certainly interested in the street layout. In the United States it had become evident that although modern techniques can create miracles they also create many negative effects, such as the high number of accidents, especially involving children. He praised the equal treatment of all transport modes and concluded that Radburn was an escape into old times, the period when there was still peace in the streets. He hoped that the Radburn layout would also have some influence on German settlement planning. In Radburn there was no conflict between the past and the present; both were in equilibrium.

Apart from Radburn, a few British and German architects were also concerned about road safety in the housing estates they designed. It is not clear whether Barry Parker used footpaths in order to reduce contacts with the modest amount of motor traffic or whether he used footpaths only as a 'gimmick' which had no significance whatsoever. The plan of New Earswick shows that the parts built between 1911 and 1922 had already some footpath connections leading to the primary school, playground and the village green (see Fig. 4.1). These footpaths would connect cul-de-sac streets with each other and were different to the footpaths in Radburn.

Creese gave the impression that Barry Parker used the Radburn layout when designing Wythenshawe (Creese 1966: 266). The quote which misled Creese was from Parker (possibly 1940). Parker wrote:

At Wythenshawe they [motor roads] bound neighbourhood units, and schools are so placed that no children cross them [the parkways] when going to school, nor need people cross them when going to the shops (Parker 1940: 9).

Clearly what Parker meant here was the crossing of parkways and not roads in general. Studying the existing plans of Wythenshawe from 1932 and 1933 it became clear that Parker did not use the Radburn layout. The earliest existing plans designed as the Benchill Estate in 1932, the Brownley Green Areas No.2 in 1932 and 1933 and the Lawton Moor Area in 1932, were largely built as they were designed. The main feature of the Wythenshawe plan was different forms of culs-de-sac; but in all the plans children had to cross roads, though not the parkways, when going to school or to any other general infrastructure facility. Apart from the parkways, some of the major roads, e.g. Wythenshawe Road and Greenwood Road (see Fig. 6.1), also did not have any houses on either side. The housing plots were set back about 45m from these roads.

There is no doubt that Parker knew about Radburn and he had seen drawings of it in 1925 (see Fig. 6.4). Parker criticized the layout on several occasions. He was of the opinion that there were still too many cul-de-sac roads leading into 'main' streets.

This indeed was a major criticism of the street plan. The street hierarchy in Radburn was not very sophisticated, at least not by German or even by British standards; it consisted primarily of two types of streets, culs-de-sac and rather wide residential streets which would seduce motorists to drive at high speed. Though the Radburn layout was safe for pedestrians it may not have been particularly safe for motorists.

However, the fundamental difference between Europe and the United States was the degree of motorization and the provision for it. In Radburn the main concern was not only to protect residents from motor traffic but also to provide plenty of road space for the car, whereas in Wythenshawe Parker's main objective was to reduce the number of main roads in order to reduce costs. Like Sunnyside, Radburn showed many similarities to German street layouts, particularly the ones designed by Hermann Jansen. It is interesting to note that in a very early version of the Radburn plan the garages were grouped together, something which was commonly used during the late 1920s and 1930s only in Germany. Later in Radburn, garages were built next to the houses. One could also argue that very similar to German planning was the design of the green space. But again the German connection can possibly only be proved with research carried out in the United States.

Yet Radburn was not a success. Its house prices had from the start been much higher than in Sunnyside. When in 1929 the Wall Street stock market collapsed, the City Housing Corporation was financially ruined. Only two superblocks were built and about 400 families moved in. Despite the partial development, Radburn showed the new design forms, and today it is still very popular as a residential area though sadly its name has disappeared. When Stein went back after twenty years only two fatal accidents were reported to him, and both were on the main highways (Stein 1958: 51). In the coming years other housing developments were influenced, designed or built by the RPAA, such as Chatham Village in Pittsburgh, the Valley Steam Project near New York and in the greenbelt towns.

The influence of the RPAA in Europe before World War II

There is little doubt that the RPAA was influenced by British and possibly too by German planning ideas. However this influence was not only one-sided. There was one issue in particular where the European planner could learn a great deal from the United States. It was how to deal with motor traffic, particularly in built-up areas. By the mid-1920s the United States already had a very high use of motor vehicles. In 1926 the number of cars per inhabitant was about 8 times higher in the United States than in Britain and 35 times higher than in Germany. The Americans practically invented traffic management which had been in use in some cities since the turn of the century. The architects of the RPAA attempted to design residential areas in order to live peacefully with the car.

Raymond Unwin, who visited the United States regularly in the latter part of his life, vividly described the traffic congestion in some of the major cities along the East Coast in 'Higher Buildings in Relation to Town Planning', written in 1923. Already he had tried to foresee the impacts motorization would have in Britain. His main point in this article was that

where traffic has already reached the comfortable capacity of streets any further increase in height must cause or increase congestion (Unwin 1923: 12).

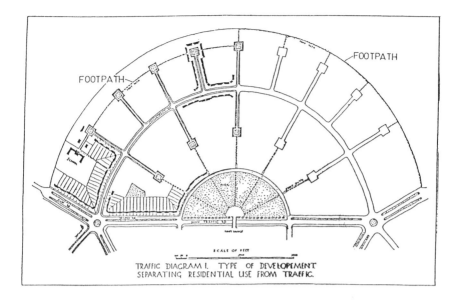

Figure 6.5 Unwin's traffic diagram of London, 1933.

He concluded 'we in Britain do not know what it means to go for total motorization' (ibid., p. 3). He also pointed out the insoluble problem of car parking in the central area of Chicago.

However the most striking impact of the RPAA and also his contacts with Thomas Adams – the General Director for Regional Planning in New York – can be seen from Unwin's work at the Greater London Regional Planning Committee which was set up in 1927 under the Minister of Health, Neville Chamberlain. Unwin was involved from the start, and he was formally appointed as Technical Adviser to the Committee in 1929. It was an advisory document on many different issues, including some remedies against ribbon developments on main roads. His ideas of main highways which should not have any adjacent buildings was a mixture of American parkways and Benton MacKaye's 'townless highways'. For him such roads could become a real cinema for car users. MacKaye used a similar picture when referring to his high-ways; for 'the benefit of the motorist the pleasant views and aspects of the country' would be obtained (MacKaye 1930: 94).

Apart from two interim reports published in 1929 and 1931 the final report, which included many issues raised in the first one, came out in March 1933.

In his chapter on 'Road Traffic' in the final report, Unwin demanded that the growth of accidents and the number of road deaths could not continue to be tolerated, a notion many of his contemporaries shared.

His traffic diagrams included the total separation of the different transport modes including a footbridge for a nearly totally independent footpath network (see Fig. 6.5). He suggested three types of road for through traffic:

1 the typical American parkway with strips of open land at various widths according to the local conditions;
2 in built-up areas, protected roads 'fenced against frontage access on both sides with development provided for on subsidiary roads to which the buildings are provided with access would be most appropriate (ibid., p. 63);

Unwin developed a third category, roads which carried heavy industrial traffic; pedestrians should not be allowed to cross such roads at all. If crossings were necessary they should have bridges or tunnels. Footpaths and in some cases cycle tracks should preferably take a separate route. Unwin was of the opinion that roads for heavy industrial traffic should be treated like railways. Here again one can find another notion MacKaye had used before, referring to his 'Townless Highways' as railroads.

We still think of it – the car – as a homely and companionable vehicle . . . when in hard fact it is a species as distinct as the locomotive. By a similar transfer of habit, we have until very recently looked upon the motor road as a fitting frontage for our home lot instead of regarding it realistically as a causeway as much shunned as a railroad (MacKaye 1930: 94).

Unwin found roads exclusively used for light passenger traffic desirable for the future. Traffic in built-up areas should be separated wherever possible. He in fact suggested – if the finance allowed – Radburn-type street layouts, though he did not mention Radburn.

In order to improve traffic flows on major roads, the number of side streets had to be reduced. He suggested side streets could be changed into either pedestrian streets, culs-de-sac or one-way streets.

Unwin's transport ideas were clearly influenced by the developments in the United States; because they were extremely advanced in comparison to his fellow countrymen perhaps it is not surprising that nothing was put into practice.

Conclusion

The United States developed significantly differently in terms of street planning and motorization compared to Britain and Germany. One problem was the traditional street layout, which was particularly prone to deform cities when motorization developed. The second aspect was the total promotion of car use over public transport during the 1920s. A further handicap was not only the weak role of planners in general, but also the even weaker role of reformers. In contrast to the British and German counterparts, it was not the 'business' of American planners to develop ideas on overall urban land use and transport policies. Connected with this aspect was the focusing of the transport planners solely on technical expertise. Thus they were not able to analyse the impact of motorization. It was the fight between the technician and the 'artistic' architect which was lost in the United States early on.

The major RPAA housing developments were influenced by British and possibly by German planning ideas but the RPAA members also influenced European planners. The combination of these different ideas resulted in the construction of Radburn, the first residential settlement which was designed with a road network intended to separate pedestrians from motor traffic. The connections between the three countries, in particular between some individual planners, were extremely close and fruitful for further work. However the Radburn layout was not adopted before World War II in Europe. At the same time, very similar ideas were developed in Germany and to a lesser extent in Britain. Some may have been influenced by Radburn but Wythenshawe, the housing estate planned and developed by Barry Parker, was not. Parker only included the American idea of parkways and the neighbourhood units in Wythenshawe but not the Radburn concept. The German planner Hermann Jansen also visited the United States and may well have been among the German delegates in New York in 1925. His street designs show a strong similarity with Radburn but his typical design was developed before Radburn and could be seen as a precursor of the German garden suburb design.

Though the Radburn layout was known by some German and British planners it was not immediately a success in Europe. My impression is that it was not so totally different from the existing examples already built in the form of British and German garden cities, suburbs and later the German *Siedlungen*. Independent footpath networks were common in Germany and also in British street layouts; though they appear to have been more usual in Germany. In addition, and this may have been much more important, since the traffic flows in residential areas, especially on residential roads, was so much lower in Europe, why should over- and underpasses be built on roads which had hardly any traffic? With the increase of motorization after World War II, the interest in using the Radburn layout rose considerably.

None of the bigger housing developments built by the RPAA was ever completed and maybe this could explain why the Radburn street layout was not a success in the United States. In fact it had less impact there than in Europe. The American circumstances made such undertakings difficult if not impossible.

The RPAA members failed ideologically in the United States because they copied too much from European planning. They were too idealistic – too European – a characteristic which has hardly ever successfully been pursued in the United States. The RPAA would have probably worked best in a German or a British context.

It is quite likely that the biggest impact the Radburn layout made in Europe before World War II was in Germany. Clarence Stein was invited to talk in front of 900 delegates at the International Housing and Town Planning Congress in Berlin in June 1931. Many of the old avant-garde were present at this conference, including Raymond Unwin who gave the opening speech. Clarence Stein illustrated Radburn. The son of Raymond Unwin, Edward, who represented RIBA, concluded about Radburn:

This is a big step in the matter of planning for the motor age and Radburn may well prove to be the basis of future planning both in America and in Europe (E. Unwin 1931: 651).

7 Motorization and street layouts during the Third Reich

Adolf Hitler and the promotion of motorization

Many biographies of Hitler stress that he was ideologically a man of the nineteenth century, but he was also regarded by most of his contemporaries as modern and forward-looking. In terms of transport (apart from a special interest in aeroplanes), he was particularly committed in pushing forward two developments: the construction of motorways and the design of a small car.

There is no doubt that the promotion of the motorway programme was the responsibility of Adolf Hitler alone. However, plans to build motorways had already been developed during the 1920s. The most influential organization was the HAFRABA (Organization for Preparing the Motor Road of the Hanse Cities – Frankfurt – Basel) which was formed in 1926. It was a private organization including members from banks, harbours, construction firms and even local authorities which had managed to build the first stretch of motorway between Cologne and Bonn during the last years of the Weimar Republic (Petsch 1976: 142). The predecessor of HAFRABA, the national Automobile Society and the Benzolvereinigung had built in Berlin a 9km-long motorway used as a race track, the AVUS (Automobil-Verkehrs- und Übungsstraße), in 1921.

The HAFRABA had sent journals and other material to the National Socialist Party since 1930. Hitler saw representatives of the HAFRABA as early as April 1933 after he had already announced the building of a motorway network two months earlier (Minuth 1983: 306). Hitler was personally very interested in the HAFRABA plans. Already in 1932, HAFRABA had suggested a 5,000–6,000km network of motorways (see Fig. 7.1). Despite opposition from the Ministry of Transport, German Rail and the Ministry of Defence, Hitler insisted on the HAFRABA plans. He extended the motorway network only slightly to 6,500km (see Fig. 7.2). A law enabling the construction of motorways came into force in June and the first stretch of motorway was opened between Frankfurt and Darmstadt in May 1935.

It was Hitler's plan to build 1,000km of motorways every year, until

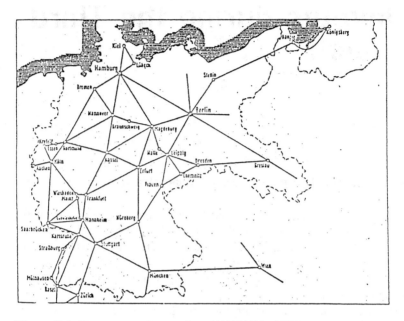

Figure 7.1 German motorway plan: HAFRABA, 1927.

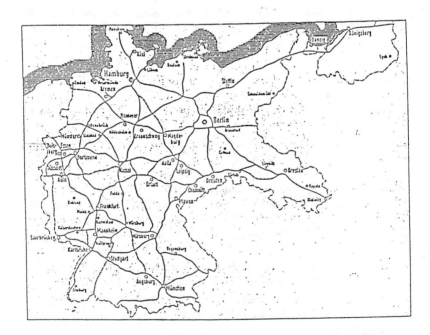

Figure 7.2 German motorway plan, 1935.

Figure 7.3 German motorways in the 1930s.
[Photographs: Sir Colin Buchanan].

Figure 7.4 British trunk road design in the 1930s. Trunk road near Dorking: the classical street layout in the 1930s was a dual carriageway with cycle and footpaths on each side. [Photograph: Sir Colin Buchanan].

the planned 6,500km were completed (Minuth 1983: 743). By the end of 1938, already 3,000km of motorway had been built. Motorways were designed at high standards. They were greatly admired abroad, especially by British transport planners. Colin Buchanan wrote from Germany in 1937, 'First taste of Autobahnen (motorways) yesterday and they are far finer than I ever imagined – I am enormously impressed' (Buchanan 1937 – postcard written to his wife) (see Fig. 7.3). Figure 7.4 shows the comparable British trunk road design. However the problem with German motorways was that there were hardly any cars to use them.

Hitler was very interested in promoting a small car, the people's car (Volkswagen) as it was later called. It is suggested that Hitler got interested in cars during his imprisonment in the Landsberg Fortress in 1923. As a prisoner he read the biography of Henry Ford and was impressed by Ford's achievement in building a cheap car for the masses (Nelson 1970: 25).

Hitler's reasons for pushing Germany into the 'motor age' were complex. It is certainly not correct to talk as Yago did 'about a community of interests between the automobile industry and the German fascists' (Yago 1984: 36). In fact, Hitler's Party, the NSDAP, was not in favour of the plans by HAFRABA and fought openly against them in the German Parliament before 1933.

There was tremendous rivalry among German car manufacturers and it was the disagreements between them which led Hitler to decide to have his people's car built by the individualist Ferdinand Porsche. Representatives of the car industry thought 'they could out-stall and out-fox Hitler' (Nelson 1970: 31, 42). According to Nelson the people's car was never intended to be used for military purposes – even the factory itself was not planned to be safe from air-raids (ibid., p. 73).

Hitler believed that the promotion of cars would weaken the existing class structure (Minuth 1983: 464). What he meant by this becomes clearer from an interview he gave to an American journalist in 1933:

That little car of his [referring to Ford's standardised car production] has done more than anything else to destroy class differences (Baynes 1942: 867).

It was his vision to make the motor car 'the people's mode of transport' (Hitler 1938 in Baynes 1942: 974). During the opening of the 24th Car Exhibition in Berlin in 1934 Hitler said:

It is a bitter feeling to know that millions of honest diligent and efficient people are cut off from the use of a transport mode which would be a source of joy especially on Sundays and public holidays (Schneider 1979: 29).

It was Hitler's aim to raise car ownership to 3–5 million in a short time period (Minuth 1983: 464). However in the next ten years he unrealistically assumed that there would be no working person who did not own a people's car (Nelson 1970: 50).

Hitler used the vision of a cheap people's car as one of his many unfulfilled promises. Clearly this idea had positive psychological effects, and so had the construction of motorways. He even talked about the psychological effect himself and its value in raising the morale of the German people. He thought that the car industry was to become the most important and most successful industry of the future, and German cars had to become the fastest and the best in the world.

In addition, Hitler also loved cars, although he never drove one himself. He had bought his first car – a Mercedes – as early as 1923 (Nelson 1970: 24) but he did not anticipate Mercedes cars for the majority of people (Baynes 1942: 975).

Hitler's promotion of motorization was one of his first priorities when he came into power. He dropped the existing car and motor-cycle tax for new registrations in April 1933 (existing cars also got some tax reductions) (Napp-Zinn 1933: 84). He also made it easier to get a driver's licence and changed some stringent traffic laws, like the abolition of the speed limit with some exceptions (see p. 43. Nelson 1970: 24).

The first 'people's car factory' was started in May 1938 but the first car did not leave the factory until 1940 (Nelson 1970: 70). By then World War II had already started and the people's cars which were produced were used for military purposes only.

118 The pedestrian and city traffic

Motorization and regional planning

The emphasis on motorization soon became apparent as a new tool in regional planning. Motorization could work against urbanization and in favour of decentralization for both housing and industry. Obviously, decentralization of industries was only feasible if a good road network was provided. It was also seen as a contribution towards narrowing the gap between the urban and the rural environment.

There was much discussion about the connection between the development of railway-dependent settlements and the direct consequence, namely the unhealthy population density in urban areas. Cars would have a decentralizing effect and could counteract such developments. It was pointed out that the car had already become an important tool in fulfilling the objective of dispersing and depopulating the large cities, and that was a precondition for the new organization of the German living space (*Lebensraum*).

Though motorization was ideologically strongly promoted it would be an oversimplification to narrow down the objectives of motorization to a 'hooray' approach for the car. The adverse effects cars could have were also understood at least in part. Particularly the planning of new settlements and new towns showed that attitude even more strongly than during the Weimar Republic. Primarily the settlement policy was a revival of traditional garden city layouts.

Relatively little is known about Hitler's attitude to the problem of motor vehicle traffic in towns. There are some indications that he was aware of the damaging effects motor traffic could have in towns, characteristically not with reference to people but to buildings. To a suggestion from the Ministry of Transport in 1933 to straighten existing roads and build by-passes instead of motorways, Hitler argued that the houses in built-up areas were already suffering from traffic and would be decaying in a few years' time (Minuth 1983: 464).

The ideal settlements: the examples of Munich and Hamburg

The promotion of housing and new settlements was one of the most important tasks of the new Government. It was part of its anti-urban ideology to create new public and private housing estates on the edge of or outside towns and cities. Some were in close proximity to industrial estates, others were planned to be surrounded by green fields or forests.

Obviously the speed of housing construction declined with the beginning of the war and stopped altogether in 1942/43 (Schneider 1979: 26). Yet the amount of new housing built during the Third Reich was nearly as much as during the Weimar Republic (1919–32: 2,036,453; 1933–9: 1,983,964 (Teut 1967: 252–3)).

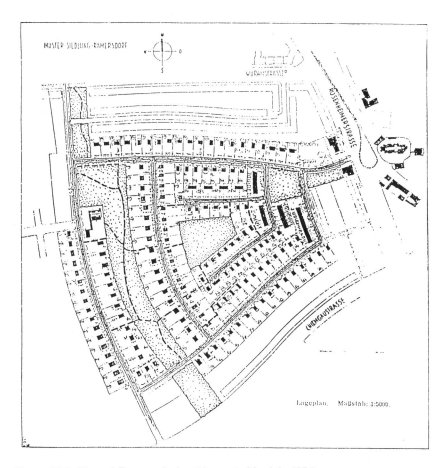

Figure 7.5 Plan of Ramersdorf settlement, Munich 1934.

In the following a few examples are given of what appear to be typical settlements built during the Nazi period. To my knowledge no research has ever been carried out in Germany or Britain to describe, analyse and classify such settlements.

Ideas of the ideal settlement and its preferred street layout were first expressed in two simultaneous exhibitions called 'The Street' and 'German Settlement' in Munich in 1934. The model settlement built for the exhibition was Ramersdorf, located south-east of Munich. It contained 192 houses of 34 different house types which varied between 60 and 120 sq.m floorspace. The houses were supposed to be owner-occupied and all had gardens between 350 and 900 sq.m. The houses were expensive, and cost between RM 8,000 and RM 20,000.

The settlement consisted of several large green spaces (see Fig. 7.5).

The main green space in the centre had the shape of a ribbon of about 450m width which acted as a connecting green space between the gardens of two rows of houses. The streets in the heart of the settlement were small (8m) and two were culs-de-sac. The settlement had footpath connections between the green spaces. No garages were built or planned. Ramersdorf also had a good public transport connection, a newly extended tramline. The criticism at the time was that the footpath connections between the green open spaces were not good enough, the streets were still too wide and one street totally unnecessary.

Ramersdorf's main characteristics were the steeply roofed houses, an extremely low density, the extensive use of green spaces and many footpath connections. This model was in the following years repeatedly planned and built all over Germany. The village green became the physical and social centre of the settlement. The movement of pedestrians was most important and the car had to play a subordinate role (Schneider 1979: 123). The street not only had functional purposes but was also seen as living space.

Ramersdorf was built as a traditional garden suburb, though the street layout had become slightly more formal than German garden suburbs had been in the past. It also included the idea of neighbourhoods which had already been applied during the 19320s. The traditional garden city ideology was intrinsically a socialist movement but the National-Socialist ideology used many of the socialist paradigms, such as the dislike for the large cities, or the concept of grouping houses around village greens. In fact, the settlements developed during the Nazi period were much more a replica of Parker and Unwin's ideal housing settlements than the actual German garden city movement was. Both the garden city ideology and the neighbourhood concept needed only some slight ideological adjustment to fit extremely well into the propaganda of the Third Reich.

Today Ramersdorf is still a very popular settlement. Its popularity stems from the green open spaces which makes it like an island in the midst of much more densely built urban surroundings. In addition the relatively large plots have given the owners many possibilities for extensions, including the construction of garages. The narrow streets allow only careful driving at low speeds. The similarity with garden suburbs built after 1900 is apparent, and also the ideas of the green centre of the 1920s spring to mind. There is one strange aspect; the lavish use of green space which appears to be in contrast to the relatively small living space provided in the houses themselves.

Apart from Ramersdorf there were in addition three large settlements planned in Munich in the late 1930s and early 40s. The most advanced one was the new South Town – Südstadt – which contained about 15,000 housing units and was planned between Ramersdorf and Giesing. The KdF Town (*Kraft durch Freude Stadt* – Strength through Joy Town), near the West Park, was somewhat smaller but planned along similar

Figure 7.6 Ramersdorf settlement, Munich, 1934.
Fountain and oak tree, two typical German symbols in
the centre of Ramersdorf.

principles. The KdF Town in Munich should not be confused with
Wolfsburg, which was called Town of the KdF Car (*Stadt des KdF
Wagens*). The last large satellite town was the North Town which was
planned near the motorway exit of Freimann, in 1942 (Rasp 1981:
183–7).

Probably the most impressive settlement built during 1936–8 can be
found in Hamburg-North, and is located in close proximity to the S-
Bahn station Kornweg. Klein-Borstel was built by the Frank brothers,
though the architect was Paul A. R. Frank, who had gained a reputation
during the 1920s for developing a particularly cheap social housing form

Figure 7.7 The centre of Ramersdorf with its large open green and the church.

in Hamburg (*Laubenganghäuser*). The Frank brothers had established their own housing corporation and the plan for the 'Frank'sche' settlement (Frank'sche Siedlung) included terrace housing for 547 families. Each house had its own garden of about 200 sq.m, so that population density was as low as 224 inhabitants per hectare.

An aerial photograph from 1931 shows that some of the area was already built upon with single family houses. In fact, the area the Frank brothers required was one already interspersed with a few houses and there were also some streets laid out. Thus the settlement was really divided into two parts which was intersected by one road and a row of housing.

However, the most remarkable aspect of this settlement is its street layout. Frank himself wrote in 1934 that a terrace housing settlement (*Kleinhaussiedlung*) did not need wide streets. Access can be given by one main road and from that several smaller roads including children's playgrounds can open up the settlement (Frank 1934: 8). He built exactly this type of street layout in Klein-Borstel. The settlement received two access roads which both end up at squares. The rest of the settlement has very narrow residential roads between 3.50m and 4m in total width. None of the houses has direct access to any road, access is given by footpaths which can be found everywhere, and these are in fact the main 'street' network (see Fig. 7.8).

The narrowness of the residential roads is strengthened by high hedges (see Figure 7.11). One of the most impressive parts is the entrance to the

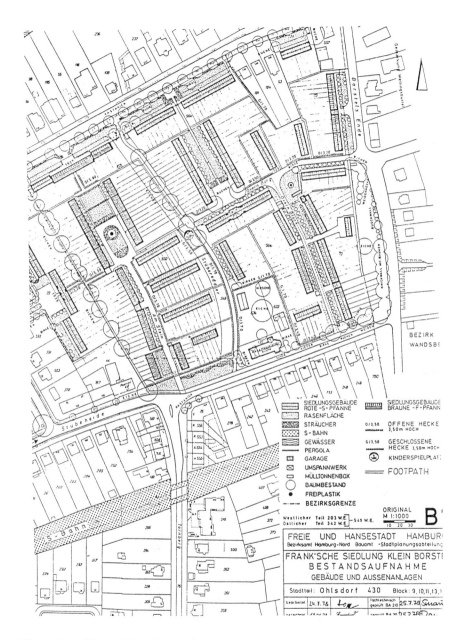

Figure 7.8 Plan of Frank'sche Siedlung, Hamburg, 1936/37.

Figure 7.9 Frank'sche Siedlung, Hamburg.

Figure 7.10 Frank'sche Siedlung, Hamburg: typical residential street, access by motor vehicle possible.

Figure 7.11 Frank'sche Siedlung, Hamburg: typical narrow residential street, access by motor vehicle not possible.

settlement, reminding the visitor of the medieval gate (see Fig. 7.9). The main wide footpath leading to an impressively designed row of houses has a slight upward slope, so that the path leads in the form of steps towards the centre. On both sides of the footpath are gardens and the houses are set back. The settlement has a unique quality which has recently been acknowledged and it is now a conservation area. The special features of its street layout were also realized in an exhibition in 1978 of traffic calming. Today the settlement appears to be one of the best 'natural' traffic-calmed areas in Hamburg. Some of the streets have, because of their narrowness, shared road space, and car drivers have very little choice other than to drive at low speed (see Fig. 7.10).

Most, possibly all, of the planned housing settlements built during the Nazi period, contained completely separate pedestrian footways though today many have disappeared because of legal problems – it was publicly owned land – and also because times changed. The use of separate footways was seen as an important part of settlement planning during the Third Reich. Reichow wanted to implement them in cities throughout Germany in the new German East (Reichow 1941: 336). Heinrich Himmler's guide lines for Planning and Design of the Cities in the East (Himmler 1942: 347–57, Allgemeine Anordnung No. 13/II) included separate footpath and bicycle networks.

In reality however most settlements contained many footways but not an overall connected footway network. Some of the settlements built during the time also included features of Hitler's reconstruction plans, especially the new plans for Berlin and the two new towns built near Braunschweig. The first new town, clumsily called Town of the Strength through Joy Car (*Stadt des Kraft durch Freude Wagens* or, in short, *Stadt des KdF-Wagens*), today known as Wolfsburg, and the second Waterstedt-Salzgitter or New Town of the Hermann Göring Industries, now called Salzgitter, will be discussed in the next section. These two examples may give a better understanding about the whole process of planning and the design standards used.

There was one other aspect which tried to minimize the adverse effects cars could have in residential areas; it was the policy for parked cars.

Garage planning

Community garages were designed according to the principle of the *Reichsgaragenordnung* (Imperial order for the use of garages) of 1939. In order to protect people from noise and pollution, cars had not to be parked along residential streets or in a public traffic space (Kautt 1983: 54). Therefore community garages were one means of avoiding the negative environmental impacts. Another more important argument was that cars were simply too expensive to be left on the street. Even costing just under RM 1,000 it meant more than half a year's salary for industrial workers in 1939. It was believed that car owners would never use cars for work trips but only for leisure purposes because cars were regarded as too expensive for everyday use. The *Reichsgaragenordnung* demanded that the distance between houses and garages should not be more than 500m.

There were few examples in which garages were actually built in front gardens. Some were concerned that garages in front gardens would not be peaceful enough for pedestrians, therefore footpaths were to be planned at the end of the main gardens.

It is interesting to note that many principles which were suggested during the Third Reich with reference to the organization of garages are today again policies which are and will continue to be pursued by present Ministries (both Land and Federal) (BMBau 1984).

Transport planning in Hitler's new towns: Wolfsburg and Salzgitter

Wolfsburg

The first new town was the result of special circumstances and had little to do with a deliberate planned settlement policy. If the German motor-car industry had not been so inflexible, Hitler would probably not have decided to build his own plant (Nelson 1970: 42). To build a town in addition to the plant was the idea of Bodo Lafferentz, an adviser to Robert Ley. Ley became the main organizer of the Volkswagen production (and was also head of the German Labour Front) (Schneider 1979: 30). Lafferentz argued that it was difficult to find qualified workers and in order to lure them away one had to offer new and attractive housing. Indeed from 1937 onwards, full employment had been established and workers were difficult to find. The overall housing stock in Germany was still in such poor condition that the offer of new housing was seen as an attractive way of getting qualified workers (Schneider 1979: 30).

The location of the car factory had to fulfil three main conditions. It had to be close to raw materials, not too close to the western borders; and to be near to transport facilities (Nelson 1970: 55).

According to Schneider, the proposed car factory had also to be in the centre of the German Reich because it was planned that the future car owners had to pick up their own cars (cash and drive). This policy was used to keep the price of the people's car down; it had to be no more than RM 1,000 according to Hitler's order (Schneider 1979: 31). After several suggestions, private land of the Count Werner von der Schulenburg was chosen, near the small town of Fallersleben, next to the main east–west canal (Mittellandkanal) and not far from Braunschweig (about 29km distance) (Nelson 1970: 55).

Albert Speer had the main planning responsibility of the town. He appointed the young unknown architect Peter Koller to design the overall town plan. Koller, who had refused to work with Albert Speer on Hitler's Berlin plan (Schneider 1979: 31) was unhappy about the assignment, but as he later said, was too young to say no. He was afraid to betray himself because he had to plan a town which was supposed to please Hitler and Speer but did not coincide with his ideas. His different design attempts for the main street layout, which finally resulted in a close copy of Speer's Berlin street design, show these conflicts (Schneider 1979: 34–5).

Koller, who had studied in Vienna and Berlin, was significantly influenced by Hermann Jansen. His urban planning concepts and his street layouts were derived from Jansen's work. Koller was also a student of Heinrich Tessenow, who had partly planned and built the German garden city, Hellerau. Koller wrote that Tessenow influenced his architectural design standards.

Ideologically he belonged to the 'Wandervogel' movement – a rambling association which had been formed in 1901 (Hartmann 1976: 131) – and to the German Self-Sufficient Settlement Movement – Kleinsiedlungsbewegung – an organization akin to the German Garden City Society. It had many supporters, e.g. Migge, Kampffmeyer, von Langen, Muthesius etc. Koller believed in self-sufficient settlements, a concept which was very close to his heart because he had lived in Berlin-Spandau on an allotment with people like Martin Wagner.

In order to understand the design of Wolfsburg it is significant to know about the ideology and school which influenced the architect and planner Peter Koller, because his background was surely symptomatic for other architects of the 1930s. It showed how strongly Koller was formed by garden city and closely related ideas which he could partly realize in Wolfsburg. These garden city ideas did not conflict with the architectural and planning ideas of Adolf Hitler and Albert Speer. One obvious explanation was that Hitler and Speer were interested in monumental buildings and street construction and hardly ever in the design or planning of settlements. The other reason was that the Third Reich ideology was, not only in its residential street layouts, close to the traditional Garden City Society itself.

Koller's main problem with the assignment of Wolfsburg was how to 'combine the world of gardens, flowers and vegetables with the world of cars' (Koller's letter 1987). Jansen's street layouts perfectly combined both worlds. Koller was aware of Radburn from the German planning literature and he even criticized the design, which may imply that Koller, like many others, could not imagine a similar degree of motorization in Germany.

The logic of the American design [Radburn] is exaggerated. It would mean to throw away the baby with the bath water to build the front of the houses towards the footpath and only the back of the houses towards the streets (Koller 1938 quoted in Schneider 1979: 47).

In 1987 Koller denied this criticism and wrote:

I have never criticised the Radburn plan. I always thought it was ideal (Koller 1987).

Several versions of the land-use plan for Wolfsburg were designed by Koller with the help of two architect friends during the end of 1937 and at the beginning of 1938. By March 1938 one of the plans had been accepted by Hitler. The town centre – the Stadtkrone – was, typically for Nazi planning, located on hilly terrain. It was surrounded by a 60–100m wide ring road. The width of the street was designed in order to cope with the predicted peak-hour motor traffic although it was assumed that workers would not use their cars for working trips. Such lack of logic was quite common during the Third Reich.

The planned housing along the ring road was supposed to be for business and office use, including some large flats. Part of the ring road was transformed into an axis which became the main shopping and business street. A second ring road followed the first one at a radius of 600–900m, and a third ring road was planned, partly overlapping the second one (see Fig. 7.12). Its function was to combine the upper, hillier part of the town with the lower part of Wolfsburg (Kautt 1983: 37).

The two planned main green corridors were to come as close as possible to the town centre. They were complemented by a system of small open spaces and pedestrian footpaths. The pedestrian street system was supposed to be kept largely independent from the main streets because, as Koller argued, the streets of Wolfsburg were to be used much more than other town streets by cars. It was therefore necessary to have an independent pedestrian network in order to secure quiet areas for the population who were in need of recreation and also for the safety of children. Housing was grouped together to form open green spaces and yards at the rear side (Koller 1939: 158). A nearly 3km-long green axis, used by pedestrians only, was planned leading from the highest southern hill towards the town centre. The reader may remember that green axes had been very popular during the Weimar Republic and that Koller's green axis had great similarity with Taut's publication *Die Stadtkrone*.

The construction of the Volkswagen factory and the town were started simultaneously in May 1938 (Koller 1940: 662) (See Fig. 7.13). The first housing settlement – Steimkerberg – was built between 1938 and 1939. It contained 438 housing units which were built partly as single, semi-detached and terraced housing (see Fig.7.14). It was a settlement for the better-off families. Steimkerberg was surrounded completely by forest, as the green spine of the settlement (Fig. 7.19).

1 STADTKRONE MIT PARTEIBAUTEN
2 HAUPTSTRASSE MIT RATHAUS
3 VERKEHRSPLÄTZE

--- FOOTPATH
BUILT UP AREA

Figure 7.12 Street and footpath plan of Wolfsburg, 1938.

Figure 7.13 Wolfsburg in 1987. In the background the Volkswagen Factory, built 1938.

Figure 7.14 Typical style of houses and street layout in Steimkerberg, Wolfsburg, built 1938.

Figure 7.15 Wolfsburg: back of neighbourhood shopping centre served by footpaths.

One traffic street gave access to the settlement. The streets inside the settlement were smaller, typical residential streets. A pedestrian network complemented these streets and gave access to the forest and to other parts of the town. The centre of the settlement was an open space, like a village common, and included U-shaped shopping facilities. These could be reached from the front by service vehicles and from the back by pedestrians (see Figs 7.15, 7.16, 7.19).

Today Steimkerberg is still one of the most popular housing areas in Wolfsburg. Being in private ownership, the appearance of the houses has changed but it still provides a good example of the settlements built during the Third Reich. It gives the feeling of an enchanted area,

Figure 7.16 Wolfsburg: front of the same shopping centre with motor vehicle access.

somewhat removed from the realities and certainly the cruelty of the time in which it was built. This aloofness still appears to be present today. At the beginning of the 1980s the district of Steimkerberg had become a conservation area.

Apart from Steimkerberg, other housing estates were completed (Kautt 1983: 115). In 1945 the town already housed more than 17,800 people of whom about 9,000 were foreigners, mainly Italians. The majority of them had to live in less glorious conditions than the people in Steimkerberg and the other estates, and indeed most of the population lived in huts and tents.

Salzgitter

The foundation of the second new town was the result not only of preparations for war but of purely economic considerations. Germany was still largely dependent on foreign iron in 1934. About 70 per cent of all the needed iron came from abroad. Clearly in a future war this dependence on foreign iron would be a great handicap. Therefore the objective was to get at least 50 per cent in home production (Schneider 1979: 57).

Geologists had found some poor and acidic iron ore near Salzgitter. This type of iron was difficult to exploit economically but experiments

Figure 7.17 Wolfsburg: access to houses by footpath.

in Britain (Corby) showed that an economic exploitation was possible. In 1937 the iron works 'Hermann Göring' were formed and the location of the factory was determined during the same year. With the construction of the factory, workers needed housing. In September 1938 about 17,000 workers already lived close to the factory mostly in huts. The construction of housing had already started in 1937, but the first planned housing settlement for 5,000 people started in May 1938. At almost the same time another large settlement had been started and planned for about 1,100 housing units. Despite these construction activities there was still no overall plan to build a town.

Figure 7.18 Wolfsburg: footpath connecting housing with the surrounding forest.

The location of a new town was decided in 1938. It was planned that the town would house 130,000 people. The target population was 300,000 (Rimpl 1939: 327).

The street layout consisted of a street axis which was a simplified copy of Hitler's famous plan for Berlin. The part of the street axis which led from the heart of the town towards the east included three roads. Two of them were used for motor vehicle traffic. The middle one was 90m wide and was to be used solely as a promenade by pedestrians (Schneider 1979: 74). At both sides of the main axis three-storey-high houses were built. The backs of the houses were located towards the streets and the front towards U-shaped large yards. The town contained traffic and residential streets which were designed as far as possible in complete separation, in order to keep the residential areas free from motor traffic noise. In addition, a separate pedestrian network was planned which was to provide connections between the town and the recreation areas in the west and south. Primary schools were designed in such a way that children did not have to cross a traffic street (Rimpl 1939: 326–8). Koller mentioned that such planning design was not uncommon at the beginning of the 1930s. As we have seen in Chapter 5 it was already used by Jansen during the 1920s. Koller designed a footpath network which would allow school children not to cross traffic streets in the plan of Zagreb in 1931. Even the main ring road was planned to be in a cutting in order to lead the pedestrian paths over it (Schneider 1979: 74).

■ FOOTPATH
E SHOPPING CENTRE
G GARAGES

Figure 7.19 Street plan of Steimkerberg, Wolfsburg, 1938.

It took until 1942 before Salzgitter received the official status of a town. Its official name was not, as often referred to in the literature, Hermann Göring Stadt, but Watestedt-Salzgitter because Hitler had been against the first name. But the importance of the town had already declined in 1940, after the German army had conquered the rich iron regions in France and Belgium. Yet in 1942, nearly as many as 110,000 people already lived in the town.

The examples of Wolfsburg and Salzgitter demonstrate that the functional division of road traffic was already very advanced. The open space and footpath planning of Wolfsburg showed a strong similarity to Hermann Jansen's planning during the 1920s. Many details hint that planners were either influenced by the Radburn layout or developed similar ideas along the same lines. Whatever the case may be, the Radburn layout was practically never mentioned. One explanation could be that an American design would have been regarded as '*art fremd*', or foreign, which did not fit into the National Socialist German culture.

A new approach to city centres

There were four main city-centre issues which were discussed during the period of National Socialism in Germany. All four were interrelated and were connected with street planning in the city centres.

Urban renewal

The National Socialist ideology was basically anti-urban, and hence the Nazis had no great interest in large cities, apart from eliminating their negative influence. Petsch and Walz pointed out that the main objective of urban renewal for the Nazis was to 'clean up' left-wing urban quarters (Petsch 1976: 193), yet the discussions about the need for urban renewal had started long before the Nazis came to power. Hamburg had began urban renewal, which was nothing more than demolition, in the city centre (Gängeviertel) from the second half of the nineteenth century. We also know of urban renewal close to Berlin's city centre (Das Scheunen-viertel) at about the same time. The Land Settlement Office of the Bavarian Ministry of Interior discussed the need for urban renewal in 1917. Urban renewal had already started in several other cities, such as Mannheim and Trier before 1933. Urban renewal was also seen as a means of improving the street network in the town centres. In every case the need for new streets was discussed, and sometimes new wide streets were built.

There was talk about urban renewal in the historic parts of the towns as one of the necessities to improve traffic conditions. There were other authors too who stressed the lack of access for motor traffic in the densely built-up areas. But urban renewal was not always used to improve traffic conditions, as the example of Freiburg showed.

During the last years of the Weimar Republic, Germany was in an economic crisis and hardly any funds were available for such projects. In 1934 about two dozen towns had applied to receive funds from the central government for urban renewal. Many of them were middle-sized towns in which the Nazis had a political stronghold. Petsch has argued that during the Third Reich urban renewal played only a marginal role (Petsch 1976: 192). In 1934 the Nazis had planned to improve 50,000 flats in old housing areas (Minuth 1983: 1056). But soon the interest and the money ran out, and hence what was left of the urban renewal pro-gramme may indeed have been increasingly restricted towards polishing housing facades and improving access for motor traffic.

Preservation of the historic town centres

Inseparable from the issue of urban renewal was the ideology of preserving the historic heritage of cities. Restorations of the historic part of Nürnberg, Kassel, Freiburg and other historic town centres were carried out. The restoration and conservation of the historic streets were part of this policy.

Joseph Schlippe, the main planner of Freiburg was determined to protect the historic character of the town and he pressed the town council to accept no further modernization of the historic town centre during the late 1920s. When the Nazis came to power, Schlippe's ideas were completely compatible with the intentions of the Nazis. He started restoring housing facades and changed them back to the originals; and he began urban renewal in the city centre. Freiburg was not an isolated example.

Maps of the town centres of Munich and Nuremberg show how little the street layout changed during the Nazi period. The example of Kassel showed that street improvements were sometimes mixed with urban renewal projects. However the urban streets outside the city centres were often widened, and sometimes valuable trees were cut down. Rasp talked about the need during the 1930s to adapt Munich's streets to the increasing motor traffic but the historic city centre was kept as far as possible intact (Mulzer 1972: 45).

Regensburg had an overall street plan from 1917 – the so-called Lasne Plan – which suggested extensive street widenings and new streets. This could have only been achieved by substantial demolition of historic buildings in the town centre. Luckily the Lasne plan was never implemented because of shortage of funds. Twenty-two years later the plans for Regensburg had changed. In 1939 a 25m-wide ring road around the historic city centre was planned which left the centre streets untouched.

Road traffic in town centres

There was a strange mixture of ideas about what to do with road traffic in the town centres. Broadly speaking one could classify the attitude towards motor vehicles in the centre into several approaches:

● There were planners who wanted to restrict motor vehicle use in the city centre. The first articles on pedestrianization were published. This movement seemed to become stronger as the years went by.
● Several references could be found in which either discussions about, or actual road closures took place. In several cases streets in the town centre were closed to motor traffic e.g. Rendsburg (1937); Kassel (1915), Lübeck.

- New plans were suggested for the area around the cathedral in Cologne which included a large square for the use of pedestrians only.

Adolf Abel was a strong advocate of pedestrianization. His model town was Venice and he believed that only a strict separation between traffic modes – pedestrian and motor traffic – would secure the survival of cities. His plan for the town centre of Wuppertal (1938) included several large pedestrian axes. The article 'The Return of Traffic Space to the Pedestrians' published in 1942, was another example of intensifying the discussion about pedestrianization in town centres. He talked about the importance of this subject in modern urban planning. A plan for Baden-Baden was included in the article which showed a large pedestrian network for the historic spa-town in 1942. The work of Adolf Abel became better known after World War II. He was not particularly popular with the Nazis after Hitler had openly disapproved of Abel's reconstruction plan of the famous Glaspalast in Munich.

According to Petsch the discussions about pedestrianization became more intense after 1942. It was not only planned to close the 'party areas' to motor traffic (these streets were used for marching and cheering the Führer) but also other streets in the city, and instead to concentrate the motor traffic along the main streets (Petsch 1976: 199). Petsch's comment does not make it clear whether increased pedestrianization was the official party line or the consensus of some urban planners. I could not find any such evidence and I tend to agree with Koller who wrote that there were so few cars that the thought of pedestrianizing streets was not a real issue for planners.

There were others planners, such as Heilig, who had a much wider concept of traffic in towns, though even he pointed out that the traders would not lose out if town centre streets were pedestrianized (Heilig 1935: 740). Heilig was against street widenings because in his opinion they did not cure the problem. After one street widening similar actions would follow. His main concern was the increase in road accidents. He suggested that urban traffic should be classified into:

- main traffic streets inside towns which should be free of buildings; this view was shared by others;
- urban streets, namely the existing street network; there, the speed should be restricted according to the geographical circumstances;
- residential streets in which only residential traffic would be allowed.

Heilig's street classification showed a close similarity to Tripp's ideas, first expressed in 1938 (see Chapter 8).

Similar thoughts were also expressed by the Deutsche Gemeindetag (German Association of Local Authorities) which wrote about town centre streets. If streets were not wide enough, they either should receive arcades, be changed into one-way streets, or be closed during the main shopping hours.

There were also the traffic engineers and planners who wanted to victimize parts of the city centre to the needs of motorization. Hitler's vigorous promotion of the car gave them new impetus. There were many articles demanding street widenings, one-way streets, and new road constructions in urban areas and city centres. In order to make space for the motor vehicle, it was suggested to abolish front gardens, street trees, and side walks (or to narrow them), or to take out trams. Many of these demands were fulfilled after World War II but not during the Nazi period.

Often transport planners who were in favour of street widening in the city centre were for a strict street division of modes in residential areas in order to avoid accidents.

Destruction of the historic town centres

The final approach was most dangerous for the historic city centres; it could be described as 'the new vision for the city centre'. It was put forward by powerful members (*Gauleiter*) of the National Socialist Party. They knew little about urban planning and cared even less about the historic heritage of German cities (Speer 1970: 314–15). Most party members modelled their newly designed town centres on Hitler's plan for Berlin. Though Hitler himself had some respect for historic buildings, in his passion for creating a capital of the 'first' German people's Reich he was quite willing to destroy large historic parts of cities. He had a similar approach for Munich, but in his plan for Nuremberg the historic town centre stayed untouched (Petsch 1976: 92).

Hitler's original intention was to reconstruct four cities (Berlin, Munich, Nuremberg and Linz) into monumental cities; later another twenty-seven cities were added (Speer 1970: 176). The change of cities was based on the '*Reichsgesetz über die Neugestaltung deutscher Städte*' (The Law to Redesign German Towns) from October 1937.

All of Hitler's city plans were based on similar design principles, consisting of two main crossing street axes on which the major public buildings were located. In the plan for Berlin and Munich a Great Hall dominated the design. The square in front of the Great Hall in Berlin was to be kept free from motor traffic, which gave Speer some problems with the north–south traffic flow (Speer 1970: 77). As all the town designs were very similar, one can assume that the feature of a traffic free main square was copied in all the other plans. But these squares may have had little to do with pedestrianization. Their main intention was to emphasize architecturally the Great Hall or some other monumental buildings of the Third Reich.

There was no clear party line on any of these approaches. All four existed, and all may have been present in some towns. Some, like urban renewal, weakened because of lack of money and resources. Maybe this pluralism could be explained by the disinterest of the Nazis in existing urban areas even after 1935. Though several authors have remarked on the change of Nazi attitudes in favour of large cities by 1935/36, it obviously did not include a clear answer to the traffic problems in the city centres. It meant little if a planner day-dreamed about a car-free city centre and the traffic engineer of the same town could politically push through some street-widening projects. Despite these ideas very little changed in most towns.

Conclusion

The Third Reich was based on ideologies which derived largely from the start of the century. They had continued subliminally during the Weimar Republic. The planning ideology was backward-looking and it would have pushed Germany back into the pre-industrial age if the strict party ideologists had fully prevailed. The recent example of Cambodia may give an idea what this may have implied. Larger-scale motorization was not originally part of the National Socialist planning ideology.

Hitler's plan to bring Germany into the 'motor age' appears to be a somewhat contradictory approach but it blended in very well with some aspects of the National Socialist ideology. He pushed forward motorization in Germany which had been rather backward in comparison with most other West European countries. He understood the significance the motor industry could have for the future economic development of Germany. The motorway network became the most advanced in Europe. Many of his ideas in connection with the car industry only became of major importance after World War II. In this respect he may have laid some of the roots of the German 'Economic Miracle' after the war.

The effects which increased motorization could have on the urban and regional structure of Germany were seen much more clearly later, in the 1950s and '60s. However, the Nazis developed no new ideas in urban or transport planning. Neither the urban renewal programme nor their settlement programme included anything new. All these concepts can be found during or even before the Weimar Republic. As seen before, the Nazis simply picked up ideas which had already been discussed or carried out in previous time periods, and made them part of their own ideology. Clearly with a centrally controlled state and a huge propaganda machine this was easily done.

New settlements were designed similar to the traditional garden suburbs. Several of the design elements, which Barry Parker and Raymond Unwin had used for British settlements, were used in Germany.

Often the German settlements were much more a copy of British garden suburbs than in previous decades. This design was seen as typically 'German' whereas the housing settlements built during the Weimar Republic, which mostly had flat roofs, were condemned as 'art fremd'. One could even argue whether details such as the main axes used in Nazi settlements or in Hitler's reconstruction plans were copies of Unwin's main streets. Another reason why garden suburb design was popular may simply have been that many active planners and architects during that time had been trained by architects who were close to the garden city ideology, even the representatives of the 'Neue Bauen'. There were few conflicts between the design of a traditional garden suburb and a suburban settlement built during the Third Reich. Apparently the Nazi period is the last period in which garden suburbs were built. With the end of World War II this tradition which started in 1902 was broken.

Some of the settlements built during the Third Reich are still good models of their time but the settlements I have described were, as Killer wrote, 'Raisins in a Cake' (Koller 1987). But many of the housing settlements were of very poor quality and had very low standards (no bathroom, very small rooms etc.) at least in comparison to British housing.

The idea of street division with reference to difference transport modes became more prominent during the Nazi period. It appears strange that a totalitarian state which had little respect for human life was concerned with pedestrian movements and safety aspects. Clearly the Nazis were haunted by the increasing number of motor vehicle accidents but this may not be a convincing enough explanation. It may have appealed to the totalitarianism of the time to have orderly separated transport modes. However to suggest that the approach to separate transport modes could be explained totally by the type of political system oversimplifies the issue. There is no clear answer to the question of why independent footpaths were important in settlement planning. It could well have been that, as with the promotion of the car, the negative impacts were seen more clearly than in the post-war period. There is also no doubt that environmental questions were part of Nazi ideology and therefore the negative environmental impacts were acknowledged.

As discussed in Chapter 6, Clarence Stein presented his Radburn layout at an international conference in Berlin in 1931, two years before the Nazis came to power. The Radburn layout was discussed much more in the German planning journals than in Britain but again we do not know how influential this layout was for residential German street design. Some details presented about Wolfsburg and Salzgitter could give the impression that it was well-known and used but one has to remember that some German planners had designed very similar street layouts to the one in Radburn since the Weimar Republic.

The approach in the city centres was blurred, and a clear party line is difficult to identify. It seems that towards the later years of the Third Reich the idea of pedestrianization became attractive and was accepted. But again there is no obvious explanation for this. Maybe it was seen as the only option to keep the historic city centre intact, particularly because in nearly every town, wide ring roads had been planned and largely built. But as mentioned above many of these plans dated back to the Weimar Republic and before.

Hitler's own ideas on city centre reconstruction would have devastated the character of German cities. In terms of traffic flows it would have generated huge amounts of motor traffic inside the cities, because of the over-dimensional street layouts. This however was incompatible with Nazi ideology.

Research on settlement and town centre planning during the time of National Socialism is still very patchy. Even today it is regarded as a sensitive subject. This chapter may have helped to understand some of the ideas but its intention was not to give a complete overview of planning during that period. Looking only at one aspect was difficult because many documents have been lost and with them the possibility to carry out more substantial research on other questions. One conclusion which may appear to be surprising is that planning in general and in particular was riddled with contradictions and logical inconsistencies during the Third Reich.

In Chapter 9, which is concerned with the developments since World War II in the Federal Republic of Germany, there is some continuation of the planning ideas developed during the Third Reich. But most of them had disappeared by the end of the 1950s. It is astonishing to find that from the 1960s onwards hardly anything remained in the memory of transport planners about the street layouts of either the Weimar Republic or the Third Reich. It was as if only a blank space remained. German planners did not look at their own history. Many clearly wanted to change their cities to cope with full motorization but funnily enough history played another trick on them.

In Chapter 8 we return to Britain, where during the 1940s the first serious large-scale road planning took place and the battle over road safety continued.

8 The British approach towards urban road transport and pedestrianization from the 1940s to the 1960s

Urban road transport policies during the 1940s and 1950s

Road transport policies in Britain during the war and post-war period have to be seen in connection with a planning euphoria, previously unknown. This period can be regarded as the heyday of urban and regional planning with the formation of a new central government department, the Ministry of Town and Country Planning in 1943. Plenty of books have been written about this period and the list of authors is too long to be included here. Many important and famous reports by committees and individuals formed the basis of the new planning legislation which contributed to fundamental changes in Britain's economic and social structure. Between 1946 and 1950, a large number of new towns (14) were designated, planned and started. Urban and regional planning strengthened as a professional and political force. It combined a whole range of different academic fields and road transport became an important part of it. New planning ideas and concepts were developed, not only for the new housing settlements and new towns. Significant changes were also suggested for the existing built-up areas. Road transport was seen as a decisive factor in helping to build a new Britain.

One of the issues of road transport policies during the 1940s and '50s was still the question of how to improve road safety, as indeed it had been in the previous two decades. It was most relevant in built-up areas and it became particularly important in relation to the increase in both car ownership and speed of motor traffic after World War II, though the number of motor vehicles only regained the 1938 level by 1948.

A new and comprehensive approach towards road building and traffic management was needed. The role of land use, urban and regional planning and their impact on traffic became an important aspect. The inadequacies of the existing urban road network were criticised by many; it was obvious that it was not able to cope satisfactorily with existing, not to mention future, traffic demands. A Report of the Departmental Committee set up by the Ministry of War Transport reacted to the growing pressure by publishing guidelines on the design and layout of roads in built-up areas in 1946. The guidelines included recommendations first made by Alker Tripp, such as those on construction and widening of radial and ring roads, and the segregation of transport modes (Ministry of War Transport 1946: 21, 26–30). Apart from traffic segregation, road safety also played an important part in this Report, and suggestions about pedestrianizing shopping streets, the construction of arcades and the implementation of precincts were discussed.

Many cities and towns made new land-use plans with reference to the guidelines of the Ministry of Transport. Some of the most famous plans were designed by Patrick Abercrombie and Thomas Sharp. All these plans advocated several ring roads (dependent on the size of the town) and substantially improved and widened arterial or sub-arterial roads. Generally the innermost ring road would run very close to the city centre. The basic idea was still to have fast and uncongested access by car to the city centre. MacKay and Cox make the point that the idea that roads should be built to accommodate motor traffic was firmly established during this period (MacKay and Cox 1979: 162). One could argue that planners only reacted to what was generally accepted in professional circles and had been advocated for half a century. In fact, Abercrombie had suggested ring roads in his plan for Sheffield as early as 1924, and again in his plan for the Bristol and Bath region in 1930.

It was only after the war, during which some British cities were badly damaged by enemy action, that these plans appeared to be realistic. The models for the road design were the United States and to some extent Germany. Little was known about the long-run social damage road building could do in built-up areas or about the fact that increased road capacity would in turn generate more traffic, though Abercrombie seems to have been slightly doubtful about the implications that main crossroads could have for the city centre. He concluded that it was not possible to forecast with certainty the effects that such roads might have (Abercrombie and Forshaw 1943: 56). Social considerations were not taken into account. This was not only valid for road planning but also for urban renewal, which began to change most major British cities.

Yet to cast these planners as an insensitive bunch of road enthusiasts would be to miss the point. The planners referred to here also saw the negative impact of motorization, and Abercrombie in particular was very much aware of it. The design of arterial and partly of sub-arterial roads would eliminate any buildings on each side; instead wide green margins were planned and often roads would have the character of parkways. These parkways were seen as the connecting link between the urban areas and the rural hinterland. The limit on parkway design was the cost. The most important justification was that these wide roads would greatly reduce both the number of accidents and the level of traffic congestion, which would in turn bring economic gains. Most planners believed in the separation of transport modes and in the neighbourhood idea – though there was a very blurred vision of what was actually meant by it.

In terms of protecting pedestrians from motor traffic several ideas were floated. Some were borrowed and had been discussed and used already in Germany or in the United States, whereas others were previously unknown and specific to Britain only.

The first concept was new. It was developed by Alker Tripp who had spent his whole professional life (45 years) working for the Metropolitan Police at Scotland Yard. In 1932 he became Assistant Commissioner and he kept the post until 1947 when he retired. Tripp's main message was stricter traffic control, more road-building and a decisive classification of traffic routes in order to protect both residential, shopping and working areas from the adverse effects of motor traffic, and pedestrians and cyclists from accidents. His most interesting idea was his demand to create precincts, later borrowed and renamed 'environmental areas' by Colin Buchanan and his team. Tripp wanted precincts for different types of activities and the name 'shopping precinct' may be the only reminder of Tripp's work today. He stressed the importance of land-use planning for creating the right conditions for a future road network. The wider issue of transport and traffic had in the past mainly been discussed by urban planners but not by people who were involved in the execution, particularly the road engineers.

The second idea which was also completely new and which would have made a considerable impact on pedestrian movements in town centres was road pricing. Adshead advocated road pricing in his book of 1941, *A New England*. He suggested three areas in Central London which would be bordered by existing main streets and a toll-gate system would be installed, in which motorists could only enter by 'special area licence' (Adshead 1941: 128). Both Adshead's concept of road pricing and Tripp's idea of precincts had in common that traffic without any destination in a clearly marked and defined area, would not be allowed to enter.

The third approach, which was partly borrowed from abroad included different methods of protecting the pedestrians directly:

- pedestrianization of shopping streets;
- pedestrian arcades; and/or
- pedestrian pavements at a higher floor level.

These issues had already been discussed by planners in many parts of the world during the late 1920s and '30s. The idea of physical separation, by which pavements would be built usually one or several floors above the roads which wheeled traffic used, was the favourite approach of several French utopian thinkers. Similar ideas were expressed in the

United States and in Germany, largely around the end of the 1920s. Pedestrian arcades were not so favourable to the pedestrian environment and could be seen more as a compromise between wheeled traffic and pedestrians. The streets had to be widened and the pedestrian pavement would be built under the existing buildings. Thomas Sharp suggested pedestrian arcades in the city centres to improve access for propelled vehicles in his book *Town and Countryside*, first published in 1932. Small-scale pedestrianization was an acceptable idea if ring and/or bypass roads were built.

The fourth concept was the Radburn road layout which had already been developed in the United States but no article can be found in Britain describing comprehensively the Radburn design before World War II. It became better known in Britain only after Stein's publications in 1949 and 1950 and with the book *Towards New Towns for America* in 1951.

In contrast the neighbourhood principle was far better known in Britain than Radburn. The American C. M. Robinson referred to it as early as 1916 in his book *City Planning with Special Reference to the Planning of Streets and Lots* which was read in Britain. In Welwyn Garden City the neighbourhood concept had been partly applied during the 1920s. The first articles on neighbourhood units as an important principle of town planning were published during the second half of the 1930s in British town-planning journals. It was believed that physical plans would produce social results and reference was made to Clarence A. Perry's article of 1924 in which Perry described the importance of specific facilities, including the street layout to create neighbourhoods. Cherry pointed out that the neighbourhood idea was discussed in a Report by a Study Group of the Ministry of Town and Country Planning in 1944, called 'Site Planning and Layout in Relation to Housing' (Cherry 1974: 131).

Three of the four concepts had some impact on post-war road planning, but road pricing was never seriously discussed until much later. In particular, Alker Tripp's ideas had substantial influence on the land-use plans of the time, yet little if anything was implemented. His major influence was on the Buchanan Report 'Traffic in Towns', published in 1963. One of the main reasons why these ideas were not put into practice has to be seen in terms of the powerful role of the traffic engineers. For centuries the role of surveyors of highways, who were the predecessors of the traffic engineers, was unchallenged. In and after 1947 with the passing of the Town and Country Planning Act, traffic engineers suddenly had as their main professional rivals county planning officers, and the planning profession was largely derived from architecture. Ideas about restricting traffic and pedestrianization came from architects/ planners – and in the case of Alker Tripp, from a policeman – but it was the traffic engineers who could put them into practice. As the rivalry was very strong between the two professions, the engineers made sure in many cases that new ideas were not implemented.

The most successful concepts were the neighbourhood units and somewhat later the Radburn road layout. The latter was at least partly implemented in the British new towns and in some new housing estates. Pedestrian precincts were built in some of the shopping centres of new towns or in some redevelopment schemes in blitzed cities but the real breakthrough came again only decades later.

The early war and post-war planned road schemes, which already included 6–8 traffic lanes, today appear brutal and had something in common with German planning of the 1930s. Politically these road-building programmes were unrealistic during a period of restrictive economic policy by the Labour party. All plans implied substantial investments on a scale possibly only comparable with projects like Paxton's Crystal Way in the Victorian period.

Alker Tripp and his first attempts at traffic calming in British towns

Alker Tripp was the first person who developed an overall approach of traffic calming in residential, shopping and working areas. His first book appeared in 1938 under the title *Road Traffic and Its Control*. A shorter and a slightly changed version of this book was published in 1942, called *Town Planning and Road Traffic*.

Tripp saw two major problems with road traffic which had to be tackled: the safety aspect and the increase in traffic volume. In his opinion the main weakness of motor traffic was that, in contrast to the development of railways, it was allowed to travel at high speed without providing sufficient safeguards for other road users.

The problem for him was how to increase the safety on roads which were basically not built for motor vehicles but for horse traffic. He concluded that roads were not yet fit for traffic, thus traffic must be made fit for roads. This implied traffic participants had to be disciplined by law and regulations until the 'right' (according to his classifications) road network was built, which was based on complete segregation of the different road users.

Tripp's views on road safety in urban areas were advanced. He wrote that

casualties could be reduced very rapidly if vehicle-speeds were heavily reduced, and the nearer the speed of cars could be brought to 3 or 4 mph the better the results (Tripp 1950: 116–17).

He pointed out that vehicle speed should be kept down at places where pedestrians and cyclists were numerous but he was in favour of allowing high speeds for motor traffic when they were excluded.

When planning new roads, Tripp argued that two types of road should be built. First, roads which acted as traffic channels, leading from one place to another; here frequency of access was to be reduced to a minimum. Second, roads for the needs of the local communities; community roads must be designed in such a way that through traffic was kept out for the safety of the people who had access to them (Tripp 1951: 40). If these two functions were not clearly divided, problems would arise in terms of reduced road safety. He used the example of Oxford Street in London, which combined both functions; it was acting as a traffic conduit and as a shopping street, with significant problems for road safety (Tripp 1950: 297).

He continued to argue that high traffic speed could only be maintained on roads which were specifically designed for it. He advocated main traffic conduits (which had the characteristics of motorways) from which pedestrians, cyclists and parked cars were excluded. Classifying roads, he suggested the following:

- *Arterial* roads for motor traffic only – motorways.
- *Sub-arterial* roads were roads for all types of traffic in which motor traffic would be predominant. Most British roads were of this type. Arterial and sub-arterial roads should be designed for major traffic flows and there, no building developments on either side should be allowed. On sub-arterial roads buildings had to be set back behind service roads.
- *Local or minor* roads would give access only to houses, shops and other premises. Here pedestrians would have priority (ibid., p. 303). Tripp was very specific about the layout of local roads, which

  must be such as will afford no short cuts to through traffic, and will not encourage high speeds (ibid., p. 310).

  These roads would only allow traffic which needed access to houses, shops and work places.

Clearly, it was easier to suggest a road network which had to be built afresh, but it was far more difficult to adapt an already existing road network to Tripp's road categories. His arterial and sub-arterial roads would have the character of bypass or ring roads. He remarked that bypass roads were often opposed by retailers because of fear of a decline in their business turnover but in reality businesses had benefited. He favoured outer and inner ring roads. The outer ring road would have the task of diverting through traffic and the inner ring road of diverting local and other traffic from the city centre. The inner ring road should either be elevated or sunken.

New arterial and sub-arterial roads should have no frontage at all, but on existing arterial and sub-arterial roads the frontage line of buildings should be set back and they should be turned around to face local roads. This way there would be no direct access from these major roads to the existing buildings (Tripp 1951: 81). If it was not possible to widen or rebuild arterial and sub-arterial roads then pedestrians would have to be isolated by fencing or put on a different level.

However his most important contribution was his concept of precincts which was to become the leading feature of the whole town plan. Precincts would consist of residential roads and were bordered by sub-arterial (in the latter publication he also included arterial) roads. Different types of precincts had to be implemented, such as shopping, working and residential precincts or precincts for historic buildings. The distance of his precincts was to be no more than half a mile at most from the arterial and sub-arterial roads (Tripp 1951: 77). Some of these precincts would be controlled by gates (here we find some of Adshead's ideas). In others the

road layout within the precinct may have to be altered in such a way as to make it deliberately obstructive. . . . The broad idea will be to give the traffic a really free run on the sub-arterials and a very slow and awkward passage if it attempts to take short cuts through the precincts (ibid., pp. 332–3).

Alker Tripp, for the first time in history, enunciated a major part of the traffic-calming concept which is today advocated even by the more conservative traffic engineers in Germany and by some British transport planners. His concept implied substantial road construction and road widening for his arterial and sub-arterial roads. He thought that a major reduction of road accidents could be achieved if sufficient roads of the right kind were to be built, implying roads which would be segregated as much as possible from buildings and the weaker traffic participants. He naively believed such a programme was a practical possibility.

However for the areas where people lived, worked and shopped the adverse effects of motor traffic would be substantially reduced. Roads in such areas would be designed in such a way as to make them deliberately obstructive in order to discourage through traffic. We know how he visualized some of the precincts. They should be like the Inns of Court in London. In his opinion,

the layout of local roads has to be devised to check any invasion by fast through traffic; deliberately 'obstructive' designs will for that purpose be not only legitimate, but desirable (ibid., p. 349).

Unfortunately he did not suggest a wide variety of traffic-calming measures but he did however argue that no treatment of the precincts should be alike and the traffic situation in the neighbouring areas had to be taken into account.

The post-war plans of Abercrombie

Tripp's idea of creating precincts in order to protect residents, shoppers and the working population from the adverse effects of motor traffic was most closely followed by Abercrombie. Not surprisingly the idea appealed to him, for he had been brought up in the garden-city tradition. All of his war and post-war plans included precincts as an important concept of town planning. Abercrombie's most well-known plans were the County of London Plan (1943) and the Greater London Plan (1944). But he was also involved in other plans, such as Plymouth and Bath.

Other planners also used the idea largely because it had been included in the manual of the Ministry of Transport in 1946. Sharp's plans were slightly different but basically he too wanted to protect some of the historic centres from through traffic, though he was mainly concerned with the preservation of historic buildings.

Generally, there was some slight confusion between the notions of precincts and neighbourhoods. The notion of neighbourhood was used for describing larger urban units. Particularly common was the concept of shopping precincts for town and city centres.

London's transport problems had not changed significantly since the turn of the century. Central London was choked with traffic and the number of accidents was high. Tripp gave examples of the growth in the number of vehicles. In 1904, 29,000 vehicles passed through Hyde Park Corner (12 hours) but in 1937, 81,000 vehicles were counted (Tripp 1950: 4). At nearly all major junctions and counting points, traffic volumes had more than doubled since 1904. In 1949, traffic had actually fallen by about 10 per cent in Inner London in comparison with 1937, but this was mainly the result of petrol rationing and it was not difficult to foresee the future demands for road space.

What had been missing in London was a substantial road building programme which was already being carried out in many other European capitals. Major roads had been built mainly during the 1920s and '30s, but they were wholly restricted to outer London. Many of them included either separate bicycle/pedestrian tracks or separate cycle ways and pedestrian footpaths (see Fig. 7.4). Patrick Abercrombie, who was well aware of the road construction abroad, developed an overall road plan. His intention was, as he said himself:

Some drastic action is necessary if any real headway is made in solving London's traffic problem (Abercrombie and Forshaw 1943: 53).

Abercrombie pointed out the main defects of London's traffic which were the mixing up of all forms of traffic, there was little distinction between through and local traffic despite the road classifications by the Department of Transport. There was also insufficient width of main roads, too frequent intersections, and there had been only piecemeal attempts to improve London's road network (Abercrombie 1945b: 65).

The County of London Plan of 1943 and the better-known Greater London Plan of 1944 included many of Alker Tripp's ideas. As suggested by Tripp and others before him, London would receive several ring and substantial arterial roads. The arterial roads had a width between 110ft (33m) and 400ft (120m); flyovers and specially designed roundabouts were planned for intersections (Abercrombie and Forshaw 1943: 55). The widths of sub-arterial roads were not given but appear to be not wider than the North Circular Road. In the Greater London Plan, Abercrombie concluded that no general directions could be given for the overall width of roads, layouts etc. (Abercrombie 1945b: 68).

Abercrombie frequently used the word 'precinct' without ever defining it. Despite that, he was concerned about the widespread invasion of precincts. He pointed out that roads of various classes carrying large volumes of through traffic had been allowed to continue, or had even been planned, to pass through areas which should be free from the danger and inconvenience of such traffic. Abercrombie suggested 5 ring roads for Greater London which he numbered from A to E. A, B and C were ring roads located inside London's Metropolitan area, though large parts of the C-ring were outside the County Council's boundaries. The A-ring, a sub-arterial road, was closest to the centre of London and would have connected all the terminal stations. This ring had been proposed before by the Bressey and Lutyens Report for Greater London in 1937 and again by the Royal Academy Planning Committee with Lutyens and Bressey as Chairman and Vice Chairman, in 1942 (Buchanan 1956: 225). The street proposal of the 1937 Report was published by the Star under the title 'The Spider Web of Greater London's New Roads' and included many features of Abercrombie's later plan.

The B-ring was to be an arterial road, but parts of it would become a parkway. The C-ring – again a sub-arterial road – would consist of the North Circular and the proposed South Circular road. The D-ring was planned as an express arterial road (motorway) and had to be newly built. Most of these planned ring roads were to incorporate parts of existing roads. The last ring, the E-ring, was designed as a parkway. About forty years later the D and E-ring became the orbital motorway, the M25.

Abercrombie also suggested a parkway for the central arterial road which would run from Swiss Cottage over the new Waterloo Bridge to the Elephant and Castle. Part of the east–west arterial road also had the character of a parkway and would partly run in a tunnel in the central parts of London. The Plan included 10 express arterial roads which ended at ring D and continued as arterial roads to ring B.

The system of arterial and ring roads would allow the planning of precincts for residential, business and industrial areas which should be 'free from the disturbance – noise, dust, danger etc. of the main route traffic' (Abercrombie and Forshaw 1943: 51; Abercrombie 1945b: 68). Abercrombie believed that the planning of precincts would channel traffic away from areas where people lived and worked, and would help greatly to reduce the number of accidents and make London a safer place.

Apart from the Greater London Plan, the idea of precincts was also used in Watson and Abercrombie's Plan for Plymouth in 1943 and in Abercrombie's plan of Bath in 1945. In Bath he recommended that

between the radials and within the ring, precincts are formed, each of which will be free from the intrusion of buses and traffic which has no business there (Abercrombie 1945a: 41).

This was also valid for Plymouth (Watson and Abercrombie 1943: 6, 81). As in the London Plan he wanted to increase pedestrian space by an independent pedestrian network and by reducing carriageways in the precincts and improving the footpaths. As with some new housing developments in London the proposed new residential area in West Twerton in Bath was closely modelled on a neighbourhood design.

Abercrombie's approach to road planning was most sensitive with reference to the urban environment. Some of his ideas, such as the roads as major axes in town centres and independent pedestrian networks, were reminiscent both of Unwin and of German planning in the 1930s, though the precinct idea was certainly something new. He was very much aware of the destructive effects of motor traffic when he wrote:

Moreover the motor car was beginning to destroy existing communities within. The danger to life and limb increased from year to year (Abercrombie 1945: 112).

Abercrombie's approach contrasted somewhat with that of Thomas Sharp who was also involved in many town plans. Sharp used the neighbourhood idea as a social form for the urban environment but he did not refer to precincts. He had one peculiarity in that he hated bicycles and called them the plague of general traffic (Sharp 1948: 92). He was in favour of substantial road-building programmes even for small towns like Oxford, Durham, Chichester, Exeter and Salisbury. In Oxford he suggested 3 ring roads; the outer ring road was more or less built according to his plan. The city centre ring was designed, running along the edge of the university park and Christ Church Meadow in order to improve access to the city centre. An inner bypass of at least 60ft (18m) was proposed for Durham's city centre in order to give good access. Similar road plans were made for Exeter in 1946. His main principle for the city centre was to get in and out quickly by motor car.

There were several other urban plans at this time, e.g. the City of Manchester Plan of 1945. The overall design layouts were very similar to the plans described above, suggesting several major ring roads and introducing the idea of precincts to protect pedestrians in the city centre. Although hardly anything was built during these first post-war years, these plans were important because they were the trendsetters for future discussions and developments in urban road construction.

Traffic in new towns

In the new towns designated between 1946 and 1955 (from Stevenage to Cumbernauld) the future level of car ownership was severely under-estimated. Even so, transport planning for the first 15 new towns was based on one principle, which admittedly was achieved to different degrees: the separation of transport modes in order to improve safety and to a limited extent to reduce the adverse effects of motor traffic. In residential areas this was achieved by having distributor roads surrounding the neighbourhoods and service roads for access to housing. In many cases schools and other facilities could be reached without crossing major distri-butor roads. The typical design in the town centres was the shopping pre-cinct, with good access by car, which included, at least in the early years, only very partial pedestrianization. Stevenage was a glorious exception.

It is interesting to note that the street designs of the first housing estates were not copies of Radburn but had a design form which was common in Germany during the 1920s and '30s. Henry Wright, one of the main planners of Radburn, mentioned when visiting Europe in 1935 the housing estate Neubühl in Zürich which had been built in 1929–32:

There are no streets parallel to the rows but instead walks run alongside the dwell-ings, which are entered from the kitchen side. The row arrangement has a distinct advantage over the Sunnyside and Radburn arrangement, since it allows a flex-ibility in block width while the relation of the entrances to the group organisation shows distinct gains (Wright 1935: 57).

He continued to point out that this type of housing block was actually the original housing block of the German architect Ernst May, and many of his buildings had no direct street frontage but were located on foot-paths (ibid., p. 87). However we know that not only Ernst May had used this type of design (see Chapters 5 and 7).

Abercrombie in the newly-drawn-up plans for a neighbourhood of 12,000 people in West Ham had most houses designed with no direct access to roads, but only footways. This type of layout was also used for the New Town, Ongar, later not designated. Some of the neighbour-hoods in Stevenage, Glenrothes, Corby, Harlow, and Basildon used a similar design (see maps in Osborn and Whittick 1969). This type of design became less popular in the later years, and if at all was found only in small parts of a neighbourhood. Thomas Sharp heavily criticized such layouts:

The street approach to houses is sometimes deliberately avoided by some modern architects and a terrace or a block of flats being set at right angles to the road and approaches by footpath only. It is argued this is done to avoid traffic noise. To cut off vehicle traffic is a very queer way of planning in a transport age (Sharp 1940: 92).

Thomas Sharp was himself appointed to design Crawley in 1946. He resigned one year later and the Master Plan was completed by Antony Minoprio (Osborn and Whittick 1969: 184) and approved in 1950. Though Sharp resigned very early, the plan of Crawley still appears to have much of his style. Sharp was an admirer of Sitte, thus hardly any of the streets were straight but all the houses face the streets. There were also many culs-de-sac, showing a similarity to Parker and Unwin's street layouts.

All residential street layouts in the first generation of new towns (including Cumbernauld) appear to be far more a mixture of German and British design styles applied in the 1920s and '30s, for instance culs-de-sac were frequently used and footway access to houses, than a copy of Radburn. The classical Radburn street layout was nowhere fully applied (with over- and underpasses for pedestrians and cyclists) simply because it may not have been very well known but also because the car-ownership level was so much lower than in the United States and was expected to remain low. Few planners had the imagination to see the drastic increase in car ownership. Even so, all new towns built between 1946 and 1960 had substantial pedestrian networks and relatively good bicycle paths because it was assumed that the majority of workers/residents would walk or cycle to work or shopping. Even Basildon, which has been pointed out in the literature as the closest copy of Radburn, did not follow the idea through in all the neighbourhoods (Dupree 1987: 104). Possibly Cumbernauld was the best example of a British copy of Radburn. Its first planning proposal is from 1958, the same year when Stein had his second edition of *Towards New Towns for America* published.

Many newly-built housing estates also attempted to apply similar design forms as the new towns. A very well-known example is Greenhill in Sheffield. However, particularly in the early examples of new housing estates and new towns, the road functions for motor traffic were not clearly enough divided and through traffic could easily penetrate the housing areas.

The physical separation between vehicles and pedestrian/cyclists faded away in the new towns built during the 1960s. Increasingly the new towns of the later generations took into account the existing and forecast car-ownership level and had less consideration for the pedestrian and the cyclist. By the end of the 1960s, the 'shared space' approach in which vehicles and pedestrians shared the same road in culs-de-sac, became a new feature. It was pioneered for the first time in Runcorn in 1966; Washington new town was another example although some parts of Washington still included an independent pedestrian footpath network.

Pedestrianization: ideas and reality during the 1940s and 1950s

Closely connected with the idea of precincts was the approach to ban motor traffic from shopping streets. Adshead mentioned as early as 1923 that certain narrow city centre streets should be converted into pedestrian use during certain hours of the day (Adshead 1923; 72). Pedestrianization schemes were mentioned in several of Ashead's later publications in 1941, 1943 and 1948. He discussed the pros and cons of different approaches to pedestrianization in *A New England* (Adshead 1941) and he concluded that the main reason for not implementing arcades was the need for too much rebuilding. He repeated his argument of 1923, that smaller streets in the central areas should be closed to traffic and converted into shopping streets for pedestrians (Adshead 1941: 131). In 1943, he wrote 'shopping streets should be designed for pedestrians only' (Adshead 1943: 18). In the plan of York which was his last contribution, published in 1948 together with Minter and Needham, several shopping streets – Shonegate and Great Shambles – in the historic city centre were to be pedestrianized and would not allow any wheeled traffic after 10a.m.

Apart from these suggestions, a limited possibility existed from the beginning of the 1930s onwards to close streets to motor vehicles. According to the Manchester Corporation Act of 1934, non-classified streets could be converted into play streets and exclude propelled vehicles (Hansard 1933–4, Local Acts Ch XCVii, pp. 43–5):

Subject to the provision of this section the Corporation may by order close to vehicular traffic any street in the city which is not a classified road for a specific period on each day or on certain days for the purpose of enabling such streets to be used as a playground for children (ibid., p. 43).

Such play streets were also common in other cities during the 1930s, and Salford was especially well-known for them (Ministry of War Transport 1946: 22). Play streets appeared to have been rather effective; for instance Salford had no fatality among schoolchildren during 1936 but it also had very few cars. In addition, Salford had also used other methods to improve road safety. e.g. erection of barriers at school entrances and teaching children how to cross roads safely. It is strange that no lesson was learned from such results and play streets were only seen as a provisional measure to counteract the lack of playgrounds in inner-city areas. Even so play streets were known until the 1960s.

Alker Tripp too suggested traffic-free shopping streets. In the existing busy shopping streets shops should have two frontages and two roads, one at the front and another at the back. The back road should be for parked and waiting vehicles and the front was to be changed into a pedestrianized street.

Pedestrian schemes were also included in city and town plans designed by Thomas Sharp. In his earlier publications, he mentioned shopping arcades to protect pedestrians from weather and traffic (Sharp 1932: 187), but in his later publications small-scale pedestrianization was proposed.

Indeed there is much to be said for some at least of a town's shopping premises being situated on a street or square which is limited to pedestrian traffic (Sharp 1940: 74).

In his later plans for Exeter (1946) and Oxford (1948) pedestrianized streets were also included. In the plan for Exeter, he pointed out that one or two pedestrian shopping streets would be good for most towns.

Gibson and Ford also included a pedestrian precinct in their city centre plan of Coventry in 1939. Ford

stressed the need for areas where pedestrians could move freely and safely and he was concerned about the problem of motor traffic and pedestrians in the shopping streets (Gregory 1973: 88).

During World War II (1941), Gibson developed a new city centre plan in which he included an even larger pedestrianized area (ibid., p. 89). But the Gibson plan was strongly opposed by city-centre traders and a new and changed development plan was approved in 1951. By then many of the original pedestrian features had already been lost. Despite that, Coventry became a model of new-style city-centre planning after the war and Colin Buchanan called it perhaps the most advanced and interesting design in Europe. It was opened in 1953.

Many plans designed during the first years after World War II contained pedestrian precincts. As described above they were part of a wider overall plan containing ring roads, precincts or neighbourhoods. Abercrombie's plan of Ongar showed a picture of a pedestrianized shopping centre.

Some influences in favour of pedestrianization in the city centres, especially for the shopping centres in the new towns, came from abroad. The reconstruction of Rotterdam with its new shopping street Lijnbaan had some effect on British planners. Lijnbaan was basically a subcentre located east of the main shopping street, Coolsingel. Van der Broek and Bakema pointed out that Lijnbaan, which was opened in 1953, had more the character of a neighbourhood than of a major shopping centre (Van der Broek and Bakema 1956: 26). They wrote:

The success of Lijnbaan can be measured by the satisfaction of the shopkeepers . . . the general attractiveness of the project to a visitor and judged in this way the Lijnbaan seems very successful and fully capable of swaying the most cautious of local authorities towards pedestrian shopping precincts (ibid., p. 26).

British planners wrote about the pedestrianization of Lijnbaan but strangely the pedestrianization of Coventry which was opened in the same year as Lijnbaan was not noticed as much in the literature.

The first city-centre plan for Stevenage in 1950 suggested a pedestrian precinct. The proposal generated criticism and the new plan of 1953 gave direct vehicle access to the shopping facilities (Balchim 1980: 271). During a public meeting, the idea of pedestrianization was reintroduced and the final decision to build a pedestrian precinct in the town centre was primarily the result of a trip by planners to see Lijnbaan. After the introduction of pedestrianization in Stevenage, its town centre became the most visited pedestrian project in Britain. Despite the success of Lijnbaan, Coventry and Stevenage, most local authorities in Britain did not move in favour of pedestrianization.

Apart from Stevenage, no other new town designated between 1946 and 1955 went for a large pedestrian scheme. However most new towns included purpose-built pedestrian schemes in later years or changed the roads into pedestrian streets. Some pedestrianization was achieved in Corby, Harlow and Crawley during the 1960s. Hemel Hempstead had a mixture of small-scale pedestrian and vehicle shopping streets. Harlow and Basildon pedestrianized their town centres by the end of the 1960s. In Cumbernauld the main shopping facility consisted of an eight-storey purpose-built shopping centre which was designed in 1955 and constructed after 1962. It was a very early enclosed mall. Somewhat similar to Le Corbusier's vision of the 'vertical city', the traffic would pass through the ground floor and the upper floors were to be used by pedestrians.

There were experiments with new road designs in order to improve road safety in residential areas and to reduce other adverse effects of motorization in the new towns and in the newly built housing estates. Coventry and Stevenage became models of the modern post-war town centre which was largely pedestrianized. Despite that the transport problems in the existing urban areas were not seriously tackled.

The Buchanan Report and its impact on urban transport planning during the 1960s

In 1951, a Conservative government took over from Labour and changed some of the overall planning policies, but in terms of road construction in built-up areas little changed, not only because of economic difficulties which were extended with the Korean War, but also because of the priority given to starting the planning and construction of motorways and trunk roads. In 1958 the first stretch of motorway opened as the Preston Bypass (M6) and two years later, 72 motorway miles had been built (M1) (Starkie 1982: 6).

In urban areas the road problems had not changed; on the contrary they had been aggravated by the increase in car ownership which had reached over 3 million in 1954 and more than doubled again by 1962. In 1959, the office of the Minister of Transport was taken by Ernest Marples, a vigorous man who had made his reputation in the Ministry of Housing by achieving the Government annual target of building 300,000 new houses.

When Marples took office he appears to have had three main objectives which he wanted to achieve as effectively as possible:

1 to accelerate the motorway and trunk road programme;
2 to reform the railway system; and
3 to improve urban transport both in terms of reducing congestion and in general to come to terms with the car (Hamer 1987: 54).

The first two objectives were relatively easy to achieve. By the mid-1960s, about 300 miles of motorways had been built and the trunk-road programme of 1,700 miles had seriously been started. It is not unimportant to know that Marples had a majority of shares in a civil engineering firm (Marples Ridgeway) which specialized in road-building (ibid., p. 50). The railway system was heavily cut back according to the recommendations of Dr Richard Beeching. He suggested the withdrawal of 400 passenger services, the closure of about 5,000 route miles and 2,000 stations in 1963. By 1966 most of Beeching's recommendations had been carried out.

In contrast to trunk roads, motorways and railways, the issue of urban transport was rather complex, not only because there was a division in the political responsibilities but also because of its nature. Marples, who had read on a flight to New York, Colin Buchanan's book *Mixed Blessing* and his Report on Piccadilly Circus, became interested in the author. Back in office, he was 'told by dismayed staff that the author was a miserable, bolshy inspector' who was at the time working in the Ministry of Housing and Local Government (Buchanan 1988). Despite Buchanan's up-to-then-inconspicuous career, he was transferred to the Ministry of Transport in 1960, and was appointed as urban road planning adviser. He could soon work independently of the Chief Engineer and his only responsibility was to a Steering Group with Sir Geoffrey Crowther as the chairman. Marples had promised Buchanan that if the Report passed the Steering Group it would get published. In order to understand fully the recommendations of the Buchanan Report, it helps to know a few details about his professional career and his personal characteristics.

Colin Buchanan had in fact written a book at about the same time as Alker Tripp's first publication appeared. Like Tripp, Buchanan was horrified by the high number of road accidents and blamed many of them on the misjudgement of drivers. He also saw the mixture of different transport modes as another main reason for the level of accidents. His unpublished book called *Let Us Take the Road* dealt with the whole spectrum of accidents, including causes, localities and nature. Being an assistant civil engineer before World War II, he had to study the sites of serious accidents to see whether road conditions were responsible. As a result of his work he became interested in behavioural aspects of road users. He bought himself a camera and took photographs from his car to study the misbehaviour of drivers.

Colin Buchanan has been both a qualified road engineer and a qualified architect/planner; thus his professional background helped him to see the conflict between providing for traffic but also, in doing so, possibly destroying the architectural value of a street. He believed in the idea of towns as the centres of 'accumulated investment of centuries' (Buchanan 1961: 322). Being an admirer of the British countryside, he feared suburban sprawl and supported the compactness of urban towns. He saw that cars were the decisive force in favour of decentralization and further growth of suburbia. If the car was not controlled, southern England would disappear within a decade and become like Los Angeles.

As a private man he has always loved nature, remote places and travel. Accessibility to sites of natural beauty was both easily achieved but also quickly disturbed by individual transport. Since he has been a keen and passionate photographer, he was quick to appreciate visual intrusion by motor vehicles in historic places.

In one of Buchanan's earliest articles of 1956, 'The Road Traffic Problem in Britain', he already described the positive and negative effects motorization could have in urban areas (Buchanan 1956). As in his unpublished book he showed a great sensitivity towards the accident issue, which he has retained all his life and has repeated in nearly all his articles and publications.

His book *Mixed Blessing* was published in 1958. By chance, a publisher had read Buchanan's Minority Report, in which he dissented from the findings of the Committee on Central London's Car Parking and suggested that he turn it into a book (Buchanan 1988). *Mixed Blessing* included many issues which were discussed later in more detail in 'Traffic in Towns'. Somewhat in contrast to Tripp and Abercrombie he did not believe that precinct planning would allow total diversion of motor traffic and he saw an important problem in providing safe routes for pedestrians from one precinct to another (Buchanan 1958: 185–6). He admitted that the precinct idea was a 'plausible theory' which could be used for small new towns. Yet to solve the problem in existing cities, particularly in the town centres, pedestrian and vehicle circulation had to be arranged at different physical levels. If comprehensive redevelopment was not achieved he considered the possibility of 'motorless zones', excluding all motor traffic, including buses which polluted the atmosphere and were making a thunderous noise.

The adverse effects of the motor car were also unsparingly pointed out, such as accidents, congestion, delay, noise, air pollution, vibration and visual intrusion. It is significant that in 1958, Buchanan was already talking about atmospheric pollution. This aspect was also included in 'Traffic in Towns' but nobody at the time saw car pollution as a health hazard. Buchanan presented his assumptions to the relevant medical bodies, receiving no attention. In a government publication four years later it was stated that 'no identifiable hazard to health exists' from air pollution by motor vehicles (Ministry of Transport 1967b: para. 16). This was the official view for many more years to come. As with many of his other environmental concerns, Buchanan was far ahead of his time but was hardly understood.

In 1960, having been just appointed as an urban road planning adviser to the government, he wrote an extremely critical article against motor traffic, 'Transport – The Crux of City Planning' (Buchanan 1960). The article dealt with a dozen major questions, such as the present difficulties of urban transport, traffic congestion, possibilities of traffic restriction, accommodation of major road works in towns etc.

He started off by saying

much of our future happiness and well being depends on the extent to which we can control the motor vehicle (Buchanan 1960: 69).

He saw the urban areas as 'becoming horrible uncivilised places under the influence of motor traffic' and he continued by emphasizing that 'our whole urban tradition' was at stake and that the 'motor vehicle operates strongly against urban quality' (ibid., p. 70).

With reference to traffic congestion in urban areas he was of the opinion that the solution was not to construct and improve a few main arterial roads or to build urban motorways. He was doubtful about such road-building and wrote

it might make matters worse by stimulating travel and building up an even worse terminal problem (ibid., p. 71).

Similar ideas were expressed in other articles written before 1963. He wrote a very prophetic sentence concerning public transport which was to become true for his own country in 1961:

The consistent lesson from other countries seems to be that where they have neglected public transport in favour of the car they have come to regret it (Buchanan 1961: 72).

Buchanan's articles imply that he was indeed very critical of urban car use and I cannot imagine any radical anti-car group today being able to use stronger words. In all of them he stressed that the pedestrians' environment was of the greatest importance for the motor age and he saw pedestrian segregation as the liberation of the pedestrian. It is hardly possible to be more radical than Buchanan when he wrote:

pedestrians should have the freedom of the city. They should be free to wander about, to sit around, to look in shop windows, to meet and gossip, to contemplate the scenery and the architecture and the history. They should be treated with dignity and only as a last resort pushed down into tunnels. Discipline is the last thing they need (Buchanan 1961: 325).

His critical views about the motor car may have been softened later through the influence of the other members of the 'Traffic in Towns' team and possibly by Marples himself.

The Report 'Traffic in Towns'

The team working on 'Traffic in Towns' was truly an interdisciplinary group, consisting of seven experts and being a mixture of architects, planners and engineers (see Fig. 8.1). The Report was written with the intention that it would be understandable for the man in the street. The Steering Group apart from Crowther comprised famous politicians and planners like T. Dan Smith (Newcastle), William Holford and Henry Weston Wells (Deputy Chairman of the Commission for the New Towns). According to Buchanan, Crowther was a big help but the other members were very busy and may not have read all the details. Buchanan described later that during the early months of 'Traffic in Towns', the Steering Group met during dinners and he alone presented the progress of the work. He wrote about these meetings:

So there I was trying to put across a new viewpoint about the environment and there were they, well fed, lolling back, cigars going, saying they could see nothing wrong with the environment and what was this chap talking about? (Buchanan 1968: 52).

But by and large the Steering Group did not interfere with the contents. The results of the Report were, as Colin Buchanan admits himself, certainly not what Marples had expected.

It is beyond the scope of this book to analyse comprehensively 'Traffic in Towns' because that would be a chapter in itself. The Report was densely written and unravelled the whole urban road-traffic problem and set out the options. These were dependent on the weight applied to each of the following three variables, namely:

- standard of the environment wanted;
- standard of the accessibility required;
- availability of financial resources.

One crucial issue of the Report was the conflict which was created by the motor car in urban areas; such as the problem of accessibility which can be achieved by the motor vehicle but at the same time the damaging effect it had on the urban environment. It was stated that either one

David Crompton	Ann Mac Ewen	Gordon Michell
in 1964 David Dallimore	Colin Buchanan	Perry Burton
Peter Hills	Geoffrey Crow	

Figure 8.1 Team from *Traffic in Towns*.

restricts the car severely or one accepts massive rebuilding which had to be carried out on a substantial scale and not as a piecemeal approach (Ministry of Transport 1963: 47). Sentences like the following gave rise to severe misunderstanding of the whole Report:

if it is indeed desired to have a great deal of traffic in urban areas in decent conditions it is likely to cost a great deal of money to make the necessary alterations (ibid., p. 45).

In the context of this book I will only discuss in more detail road construction and restriction of motor traffic. With reference to road-building there was no doubt when reading the literature of the early 1960s that it was desired to have a great deal of motor traffic in urban areas. Thus the conclusion for many was to rebuild Britain's cities on a large scale. The need for road-building in urban areas was at the time not questioned by anybody. As we have seen in the previous chapters, Britain had been lacking substantial and comprehensive road-building programmes. The strongest outcry for new roads came after World War

II. This was clearly connected with both the increase in motorization and the political strength of the road lobby. Whereas the early plans during the 1940s and '50s showed no consideration for the social effects such road programmes could have on the urban fabric, the Buchanan Report was probably the first of its kind to point these out. Colin Buchanan had already mentioned these effects in earlier publications.

The Report considered the size and the geography of the city as an important factor governing how much the urban environment had to be changed in order to accommodate full car ownership. It was made clear that in large cities, such as in the chosen examples of Leeds or London, this was not a realistic option. But even in those examples three options were given – for a minimum, medium or large-scale adoption of the motor car.

Although ideas on restriction of motor traffic in the town centres or residential areas had already been discussed in Britain before 1963, it was the first time that the different options of restrictions were coupled with the pleasantness, the quietness, the level of air pollution etc. in an urban area and its accessibility by the motor car. If accessibility by car was increased, the environmental conditions of an area would deteriorate if no substantial rebuilding was carried out to accommodate such traffic.

The most important element of the Report was the conclusion that environmental areas had to be designated in residential areas in order to combat motor vehicle traffic. Many elements of Tripp's precincts acted as a model. They were compared with the rooms of a hospital or with 'urban rooms'. As Tripp and Abercrombie had imagined, the environmental areas would be of different character and the level of traffic would vary according to their function (residential, shopping and industrial). In there, the overriding consideration was given to road safety. Roads would not only be judged by their capacity for carrying traffic flows but also in terms of environmental capacity, which would be different for residential roads and main distributors. The environmental capacity as a completely new measure for roads would provide both the standards and limits for the environmental areas. That was truly something nobody had ever heard of before.

It was pointed out that there was some elasticity in the environmental capacity of roads and a metaphor was used to explain it; just as dwellings had some elasticity in the number of people they could house, there is certainly a limit, otherwise they will become a slum.

In terms of protecting the pedestrian, Buchanan's team favoured in some conditions, such as high density of both pedestrians and motor traffic, a complete segregation of one from the other, but in different conditions, a mixture of pedestrians and vehicles was not seriously harmful if vehicles were to reduce their speed and volume. For shopping areas, pedestrianization was seen as a good concept.

There was another aspect which should be mentioned here because it echoed the thought Raymond Unwin had first expressed in his article 'Higher Building in Relation to Town Planning' (Unwin 1923). Unwin had pointed out, in the context of the United States, the relationship between the floor space of buildings and the required road space to accommodate cars. Buchanan's team continued with this idea and discussed the issue of town traffic as a function of buildings in some detail (Ministry of Transport 1963: 80–3).

Colin Buchanan and his team did not offer attractive answers when they concluded: 'There is no easy and complete solution to the problem which developed from the growth of motor traffic' (ibid., p. 23).

The impact of the Buchanan Report

Since the publication of 'Traffic in Towns' on 27 November 1963, much has been written interpreting the contents and the philosophy of the Buchanan Report. When the Report was published it was widely reported in the press, on radio and television but it did not create such a stir as the Beeching Report which had also come out during the same year. The Buchanan Report received little criticism, apart from *The Economist* which wrote an attacking article under the headlines 'Traffic without Economics'. The Government itself accepted Buchanan's conclusions in principle, though later publications made it clear that there were some reservations about them.

About fourteen days after the Report was published the British Road Federation organized a conference under the title 'People and Cities'. Buchanan made some slightly cynical remarks about the interest the members of the British Road Federation had in people and cities but concluded that they must now be 'passionately concerned' about this issue otherwise they would have not organized the conference (Buchanan 1963). The controversy about the Report became already apparent at this meeting. However the main question remained unanswered. Was the Report a surrender to the motor car or an attack on it? Marples interpreted the Buchanan Report in his own way which was the start of being misunderstood (according to Sir Colin Buchanan he did not fully understand the Report); he pointed out:

It is fundamental to the whole report that it accepts the motor vehicle as a brilliant and beneficial invention. It is in no sense restricting the motor car. All it says is that we must use our motor cars to the maximum, and yet be sensible and keep some good environmental areas. We have to face the fact . . . that the way we have built our towns is entirely the wrong way for motor traffic. We want an entirely different type of town (Marples 1963: 12).

Geoffrey Crowther commented that people will bitterly resist even the most necessary restrictions of car use. He warned that we may fall into the same trap as the United States where freedom of the car is everything and 'then you have as much environmental standards as there is left over' (Crowther 1963: 25).

Conflicts between Dr R. J. Smeed, Deputy Director of the TRRL, and Buchanan were looming when Smeed argued that there was no need to be so pessimistic about accommodating private cars and he had several suggestions on how to improve road capacities, e.g. change of vehicle occupancy, staggering of peak hours and smaller cars (Smeed 1963). Smeed was at the time already involved in writing a report for the Ministry of Transport on road pricing, researching its economic and technical possibilities and this was published in 1964 (Ministry of Transport 1964).

In discussing the impacts of the Buchanan Report had on Britain, it may be useful to distinguish between the more general impacts, such as the impact on the planning system, on land use and transportation studies, and the more specific ones, such as the encouragement to pedestrianize shopping streets or the designation of environmental areas. The latter will be primarily discussed in Chapter 11.

Impacts on specific towns

As Buchanan has pointed out, when writing the Report, there had already been some local authorities which had been interested in implementing Buchanan's ideas. There were firstly the chosen cities in the Report but also other local authorities, which wanted to get advice from Buchanan. After the Report was published and received not only its official blessing but also great publicity, local authorities became increasingly aware of the great need for a new approach towards transport and land-use planning. By then Colin Buchanan had left the Ministry of Transport and taken up a chair at Imperial College, London (October 1963). He had started a couple of months later his own consultancy (Colin Buchanan and Partners). The demand for studies carried out by the new firm was so high that they could not cope. The first study carried out was *Bath, a Planning and Transport Study* (1965) which was the successor of an earlier Report 'Traffic in Bath' (1964). Another large study dealt with Cardiff (1966 and 1968), but there were many more to come, such as the transport studies in Edinburgh, Sheffield and Tyne and Wear.

In the study of Bath, the objective was stated as

bringing under control the anti-environmental effects of motor traffic . . . We . . . do not conceal our view that the motor vehicle has in many ways made itself a serious and major social nuisance (ibid., p. 8).

The methodology was similar to the case studies in 'Traffic in Towns',

Primary road		Local distribution road in·cutting	— —
Primary road in tunnel			
District distribution roads		Pedestrian areas and footpaths	
Local distribution roads	—	Car parks ⊛ Depots ▨	

Figure 8.2 Pedestrian plan of Bath by Colin Buchanan and Partners, 1965.

though it was clearly more detailed. In total it showed great sensitivity towards conservation and pedestrian movements, considering that a high car ownership level was assumed to be 440 per 1,000 population, and anticipating a significant population and employment growth by the year 2,000 (Buchanan and Partners 1965).

Potential environmental areas were designated and conflicts between main traffic routes and these areas were pointed out. The suggested road scheme in Bath included two controversial aspects, the suggestion of a four-lane tunnel as a main through road at the level of the south end of Crescent Gardens, leading roughly along George Street and coming on surface level again just before crossing the river, and a two-lane distributor road was supposed to be built in a cutting under New Bond Street. The city-centre scheme provided a large pedestrian area, comparable to the scale of German pedestrianization today (see Fig. 8.2).

A similar approach to Bath was given in the Cardiff Study and in other studies which followed. In all these studies the environmental aspect played an important role. Staff at Imperial College, such as Dennis Gilbert and D. H. Crompton, started to work on models to calculate the environmental capacity of areas (pedestrian delay, level of noise and air pollution etc.). The environmental aspect of road planning was emphasized quite strongly by some planners, such as Walter Bor, who wrote:

I consider a good urban environment to be of such fundamental importance that it should be regarded as fixed and inviolate (Bor 1963: 313).

Pedestrianization as a result of the Buchanan Report

The Buchanan Report encouraged at least some local authorities to start thinking about pedestrianization and a few bold ones actually applied it. It is no accident that Norwich, which was one of the case studies in the Buchanan Report, was the first town to close a street to motor traffic, in contrast to pedestrian precincts which were purpose-built, as in the new towns.

The Draft Urban Plan of Norwich had pointed out that for the well-being of the city as a regional shopping centre, pedestrianization was crucial. It was also seen as the key policy for preserving the historic character of the city. Two large pedestrian areas were suggested in the city centre, a quite dramatic approach for its time. Alfred Wood, who was the City Planning Officer, wrote with reference to pedestrianization 'what can be done in Europe should also be possible here' (Wood 1967: 40). The truth was that most did not dare to go down the Central European road.

Norwich, like many other historic towns, had already several small shopping streets and alleys which were used by pedestrians only. In 1967, the main shopping street was closed to traffic after the Government had changed the legislation. Before 1967 the conversion of highways was not permitted. It became possible with the Road Traffic Regulation Act of 1967 (Roberts 1981: 18). A few other towns followed but pedestrianization was only carried out on a very modest scale.

There were plenty of suggestions and plans made which included large designed pedestrianized areas, for instance the Manchester Central Redevelopment Plan. Liverpool also had a comprehensive plan for a pedestrian network inside the inner motorway. There were some towns and cities, such as Reading (1968) and Leeds (1970) which implemented such schemes relatively early. Leeds, in consultation with the Ministry of Transport and the Ministry of Housing and Local Government, had published an overall concept of planning and transport, the so-called Leeds approach. It included a plan of comprehensive pedestrianization in the city centre, and about half of the pedestrianized routes were to be upper-level pedestrian ways. The publication referred to Essen as the desired example of pedestrianization which did not however have or plan upper-level walkways.

In many of these streets bus access was allowed and necessary. However as a result of the high flow of buses, the created environment for shoppers was often not particularly attractive, e.g. Oxford, Reading, Leeds, Manchester. It was later, during the mid–1970s, that the number of pedestrianization schemes increased. There was a kind of vagueness on the part of local government about pedestrianization schemes; today hardly any of the local government officers know when their first scheme was implemented. Because of this, there are different dates when streets were pedestrianized in the existing literature.

Relatively little research was carried out on the effects of pedestrianization, and certainly not by the Government's research body, the TRRL. Pedestrianization was not seen as a basic need for people, shoppers, employees but as an optional addition, which was made possible when city-centre ring roads had been built or other roads provided. None of these early pedestrianization schemes had the character of a pedestrian network.

The town centres of the new towns designated during the 1960s continued in the tradition of Cumbernauld. Most town centres received modern shopping centres, covering several floors. Runcorn's shopping city, an early example of a multi-levelled covered shopping centre, became well-known, with access by car and bus at ground level. However a number of new towns which had established town or village shopping streets incorporated the existing historic street network into the shopping areas. Most of these streets were pedestrianized during the 1970s or 1980s but nearly all of them also built modern shopping centres with a generous availability of car-parking spaces.

Traffic-restraint policies, including environmental areas were also discussed. In Newcastle upon Tyne 'environmental areas' were designated and the Rye Hill 'revitalization areas' included many interesting motor traffic restraining features, such as a pedestrian spine and the closing of back streets and inclusion of playgrounds, benches and tree planting. There must have been many more towns and cities which had pinpointed environmental areas. The astonishing story was how little was made out of it. The idea of environmental areas was further developed during the early 1970s and will be discussed in Chapter 11.

General impacts: planning structure

The Buchanan Report had the most powerful general impact on the overall planning, structure of land use and transportation planning which was seen again as a unity, as it had been during the days of Raymond Unwin.

Yet, the culmination of Buchanan's demand to combine both land-use and transport planning came much later, when in 1970 'traffic as a function of land use' was taken seriously by the Government in joining the two departments, the Ministry of Housing and Local Government together with the Ministry of Transport, into the new Department of the Environment. The combination of the two ministries was a long and painful process which started to emerge after 1963. However the main emphasis in the Ministry of Transport did not change even when Labour came into power again. It moved more and more towards promoting private motor transport. Buchanan wrote in 1965 that 'the Government is very far from discouraging the motor vehicle' by condoning the construction of motorways:

fostering the motor industry . . . [and in fact] setting the seal irrevocably on the motor vehicle as the basic transport system of this country (Buchanan 1965).

Some achievements were made during this period in terms of road safety, such as the introduction of the 70mph speed limit on motorways which was part of the Road Safety Act of 1967. It also included the possibility of prosecuting drivers above a defined level of blood-alcohol, the introduction of compulsory seat belt fittings and annual testing of cars older than four years (Plowden 1971: 356).

Land use and transportation studies

Superficially it appeared that a new form of thinking was created with the phrase 'traffic as a function of land use' but it was little more than a new slogan. This became most apparent when, towards the beginning of the 1960s, the well-known and ill-famed land-use/transportation studies were set up, such as the West Midlands Study, Greater Manchester

Study (Selnec), Greater Glasgow Transportation Study etc. The basis of these studies had been extensive traffic surveys, such as the London Traffic Survey in 1962, which was carried out by a British–American consultancy. In fact, it was the American skill of computer forecasting which was a totally new technique and could not be carried out by British experts. Yet, the time lapse between the actual collection of data and the conclusions was substantial and gave rise to criticism. These studies were all based on forecasts of high population and employment growth, connected with substantial increase in car ownership. They embodied little if anything of Buchanan's thought; they were dominated by the transport engineers, who in turn had been trained by engineers and computer experts from the United States.

According to Buchanan, the Ministry of Transport required transport consultants with experience of modelling to be employed alongside the planning consultant in the urban planning contracts commissioned or grant-aided by the Ministry of Housing and Local Government. Buchanan thinks that in certain cases this practice resulted in over-estimates of future traffic growth leading to over-elaborate road proposals.

Modelling appeared to be the answer to many problems and it was also eagerly picked up by land-use planners. They were technically skilled and were masters of modelling techniques, but they had lost, maybe unconsciously, for the first time most of their social objectives, and had little in common with the visionary planners of the turn of the century. Their counterparts, the traffic engineers, were more concerned with building roads, and details like environmental areas could be considered once all the roads were built.

Road pricing

Though the debate on road pricing had started independently of the Buchanan Report, the publication and the widespread discussions of it certainly had some impact. Smeed and Roth had suggested the application of market forces for urban roads as early as 1960. C. D. Foster had discussed the possibility of putting meters under cars which could be actuated by electrical signals and would vary from time to time and from street to street (Foster 1962). In 1962 Beesley and Roth pointed out that the freedom to buy and run a vehicle was essentially to be preserved and congestion could significantly be reduced by making better use of existing roads by various methods of restraint, or by building new ones (Beesley and Roth 1962). After 1963 the Buchanan Report was attacked by several economists for its uneconomic approach. Beesley and Roth argued that the Buchanan Report had not calculated any economic and other benefits or disbenefits of road users. They pointed out that even environmental benefits or disbenefits could be valued in money terms. If residents were to lose out environmentally then it was reasonable to compensate these environmental losers (Beesley and Kain 1964). For them, road pricing was the key to traffic restraint.

There was great controversy between the Buchanan supporters and some economists. In fact, the controversy seems not to have been as severe as was made out at the time. Beesley and Roth's article of 1962 sounds very similar to some arguments in the Buchanan Report. The problem was that the economic approach to valuing benefits and dis-benefits was theoretically sound but impossible to implement in practice as later public inquiries showed, e.g. the Layfield Inquiry.

The Government promoted several reports on road pricing and traffic restraint. The conclusion of the Report 'Road Pricing: The Economic and Technical Possibilities' was that the metering of cars as a direct pricing system was the best form of road pricing after having considered a daily licencing and parking tax. It was calculated that between £100 and £150 million could be saved under the traffic conditions of the time (Ministry of Transport 1964). Three years later the Labour Government published another report on the same issue, called 'Better Use of Town Roads': a study on the possibilities of traffic restraint on urban roads (Ministry of Transport 1967b). The report included a substantial discussion on road pricing, with particular reference to Central London, and concluded that road pricing was the most efficient means of restraint but further research was needed (Ministry of Transport 1967b). A very similar argument was used about twenty-two years later by the Secretary of State for Transport, Paul Channon, in 1989.

Apart from road pricing, other methods of restraint were discussed, such as parking control. But all this was more of an academic exercise than a practical suggestion for the future. Possibly everybody in the Ministry of Transport was too afraid of the political consequences to put these thoughts into practice, even if only on an experimental basis.

Conclusion

The period during and after World War II was characterized by a tremendous vitality in planning both in land use and transport. The new era of hope and idealism was also expressed by grand road plans and substantial rebuilding programmes in urban areas. Naturally, it included the associated traffic issues concerning how to tackle both the safety aspect and the increase in traffic flows. Many transport experts were still haunted by the large number of road accidents. The pedestrian played an important part in these plans, largely because car ownership was low and most people walked and knew what it meant being pestered and endangered by cars. It is therefore not astonishing that physical separation of motor vehicles and pedestrians/cyclists was increasingly demanded. However the economic conditions of the day made the road plans financially unrealistic, partly because priority at the time was given to new housing.

Alker Tripp developed the precinct idea which had some degree of similarity with the neighbourhood units developed in Radburn, though Tripp's precincts were much smaller in size. Precincts were later interpreted as totally traffic-free areas but Tripp had envisaged a variety of precincts, some to be traffic-free but the majority not. Alker Tripp can be regarded as the person who first included many of the modern traffic-calming policies in his overall concept of road networks. But this type of traffic calming was based on substantial road building.

Ideas on pedestrianization were discussed by a few planners during the 1940s and '50s but relatively few were put into practice. Coventry and Stevenage were the only two significant examples which pioneered pedestrianization. The closing of traditional high streets was possible after a change in the legislation in 1967. Alfred Wood, the Town Planner of Norwich, put this concept into practice for the first time. Many other cities developed plans, which included large-scale pedestrianization but the implementation was difficult because it was conditional on first building inner-ring roads.

Some of the first generation of new towns experimented with a mixture of German and British street designs known from the 1920s and '30s and the Radburn layout. Cumbernauld was the last new town which went for this approach on a bigger scale. This type of street layout was also popular in many new housing estates built during the 1950s and early 1960s but nowhere was the Radburn principle applied fully.

During the late 1950s, motorway construction started and considerably more trunk roads were built, but the transport issue in the urban areas was not tackled. In the early 1960s Ernest Marples, the Minister of Transport, appointed Colin Buchanan to study the transport problems in urban areas. The Report pointed out both the positive and adverse effects of the motor vehicle. Its importance lay not in describing the positive effects but in its warnings about the negative effects car use could have in the existing urban environment. The dilemma could only be overcome if the urban environment was adjusted to the motor vehicle, which implied either the rebuilding of urban areas at enormous cost or some form of traffic restraint. The idea of possible damage to the urban environment by motorization was something nobody had expressed so mercilessly before. It was a further development of the first thoughts expressed by planners close to the garden-city ideology.

The conclusions of the Buchanan Report were complex and did not offer easy answers. It accepted the motor car as a valuable means of transport which was certain to stay. The problem was to know what possible impacts the car would have on the urban environment. 'Traffic in Towns' set out the options. It tried to point out the facts and stated what would happen to the existing urban environment if motor vehicle use was neither controlled nor parts of the cities rebuilt.

It may have been that few planners and transport experts wanted to know because in their desperation they needed quick solutions. There were at that time severe traffic jams which seemed to get worse every day; there was the forecast of population and employment growth, and connected with that was the prediction of an even higher car-ownership level; there were accidents, and the complaints of residents about noise and fumes but there was above all the desire to own and drive a car. What else could they do but pick out the practical conclusions of the Report, overlooking the fact that they were only part of the answer? Substantial investment in public transport and the very important aspect of finding a practical measure of environmental capacity was not only far too complicated, but also something which should have come from the Government.

The Minister of Transport, Ernest Marples, obviously saw the Report as favouring maximum car use in urban areas and as a charter for the rebuilding of cities. Strangely enough, the Report was most misunderstood in the country it was written for. Whereas the German transport planners even in the 1980s see Colin Buchanan as the 'father' of traffic calming, most British planners saw in him the promoter of large, urban road-building programmes. He was heavily attacked on this issue. Hamer wrote only recently (1987):

The main theme of the Report was segregating motor traffic and pedestrians. Even now, apologists point to the more enlightened passages, such as the concept of environmental areas where traffic would be banned, or the finding that it was physically impossible to build enough roads to cater for all the traffic. but in practice it was a charter for urban road building, on an impossibly expensive scale (Hamer 1987).

There can only be speculation as to why the Germans interpreted 'Traffic in Towns' so differently. One explanation could be that by 1964 their towns were already rebuilt, and nobody of sound mind would have advocated a new wave of rebuilding. The other fact was that the Germans have always been in favour of substantial road-building. They did not make elaborate plans and discussed them in great length, they simply built them.

In 1988, the Report is as valid as it was in 1963. Sadly the urban environment has deteriorated as Buchanan and his team had predicted. The Report was far in advance of its time and was possibly written twenty-five years too early. Certainly the case studies of the report had some shortcomings which lay largely in the over-estimation by the Registrar General of future population and hence in a very high forecast of car ownership.

My criticism of the Report would be that the potential for restraining the motor car was largely under-estimated, but that was something which we may only be able to say today after decades of experience.

Mostly the lavish illustrations in the Buchanan Report implied that if British cities were considerably rebuilt then present and future traffic problems could be solved. Buchanan later admitted that this was the failure of the Report because it gave the impression of containing positive recommendations for an early and very elaborate reconstruction of towns and cities at vast expense.

There is no doubt that the Report suggested road-building but it was to be done in a sensitive way. The problem was that most transport engineers who could have put these ideas into practice did not have this sensitivity. How could they, they had never been trained for it? If the Buchanan Report had been implemented, misunderstood as it was, British cities would have been substantially redesigned and rebuilt in order to cope with the motor car, and during this process we would have lost most of what British cities stood for. In fact, the cities which pushed through large road plans and started building them were exactly lacking such environmental sensitivity, and it was no wonder that people started to protest against road construction. The further development in the fight by the pedestrian and the resident for a better environment in Britain will be discussed in Chapter 11.

9 The development of transport policies in the Federal Republic of Germany: the early years and the road-building mania

The missed chance of reconstruction

At the end of World War II there was hunger, destruction and confusion in Germany. Most German cities, particularly the large ones, were not only badly damaged but some parts had been totally flattened during the bombing raids. The period up to 1950 could be classified as the time of improvisation. It was most important to repair and patch up. People organized food and went slowly back to their work places, if they still existed, or tried to find new work. Having to break completely with the political past, nothing would have been more natural than to start planning from scratch, implementing new planning and transport ideas in new dimensions.

Many German cities started to plan along those lines. We know of three conflicting plans in Berlin: The Collective Plan, the Zehlendorf Plan and the Hermsdorfer Plan. All three had already been presented in 1945 or 1946. The Collective Plan, designed by Scharoun, was to change Berlin into a linear city, whereas the Zehlendorf Plan by Moest and Görgen was modelled on the Speer Plan. The Hermsdorfer Plan had a garden-city layout suggested by Heyer and was the least realistic one. In the end, the Zehlendorf Plan, which was the most compromising one, survived for West Berlin.

Munich had several different plans and again the most compromising one succeeded. According to Mulzer, a very comprehensive new plan was designed for Nuremberg which was not implemented (Mulzer 1972). We know of new plans for Hanover, Cologne, Stuttgart, Karlsruhe, Münster, Freiburg and Hamburg. Again, most of them were too 'modern', too innovative to be realistic. Nearly all German cities went the way of the big compromise, 'no experiment was the motto' as Alexander Mitscherlich, one of the most famous critics of post-war Germany, emphasized (Mitscherlich 1969).

The currency reform of June 1948 brought considerably more hardship in the short run which was eased with the advent of the American Marshall Plan. The Federal Republic of Germany was formed one year later and with it more overall planning responsibilities were given back to local authorities. However, it took some time before new legal requirements were passed and put into practice. The 'rebuilding law' (*Aufbaugesetz*) in 1949 eased the legal difficulties of reconstruction. But Bavaria had no 'rebuilding law' at all and because of that, Bavarian cities could only with great legal difficulty be rebuilt any differently from before the war, because of the pre-war building regulations which were still in force.

The lack of implementation of any grand overall planning ideas can be largely explained by an ideological insecurity, the unwillingness to change the existing property laws, the lack of funds and also by the pressing need to build housing at a fast rate. This became even more apparent with the waves of refugees which flooded into West Germany in the late 1940s.

The number of refugees posed such a severe problem that there was little approval by planning authorities of what was actually privately built. It is said that in Berlin only 5 per cent of all newly constructed buildings were approved by the city planning officials during these first pre-war years. This may have been true for many towns and cities. Hajdu pointed out that

a large proportion of the rebuilding was little more than a crude restoration of the pre-war urban fabric (Hajdu 1983: 17).

According to the first official statistics there was a shortage of 6 million housing units in 1950. In reality the shortage was much higher, particularly just after 1945. In addition about 13 million refugees from the East had to be housed.

Another reason why cities were reconstructed as they were was the legal position concerning property. The rights of Jewish landowners had to be re-established and the property of the department stores and other shops in the city centres belonged largely to Jewish families. The change from private into communal property, which was seriously suggested by the Social Democratic Party (SPD), was politically not realistic. Mitscherlich claimed that it was the 'taboo' of the legal property conditions which made it impossible to create new cities.

There was also the issue of underground and surface infrastructure. Whereas the surface infrastructure was often badly damaged, the underground facilities had survived the war much better. In any event, it was easier and quicker to repair the existing infrastructure than to build from scratch.

Figure 9.1 Kassel city centre – a model of reconstruction of post-war Germany. Main pedestrianized shopping street with tram and bus access.

Finally, Germany has always had a strong sense for its urban heritage which can be traced back to the medieval times when the 'free city' (freie Reichsstadt) had political and psychological functions for its inhabitants similar to a state. Such unconscious feelings must have clearly been strengthened when citizens saw their towns in ruins.

All these factors together had the result that most war-damaged German cities were largely reconstructed as they had been before. Mitscherlich pointed out that after the was the Germans had missed the opportunity to build more intelligently thought-out new cities (Mitscherlich 1969). Hillebrecht also talked about the missed chance. Instead of

Figure 9.2 Kassel city centre: Treppenstraße ('step' street).

clinging to traditional forms, leafy and generously spaced cities should have been built with satellite towns at the edges of the urban areas (Hillebrecht 1957). Many other planners were of the same opinion. This was particularly valid for the city centres though there were differences.

There were hardly any major towns which went for a totally new approach in the city centre with reference to the street layout and the transportation network; Kassel was an exception. In 1950, its land-use plan was passed, which had been the result of a nationwide architectural competition. In this plan, the medieval city centre street layout had nearly disappeared and was replaced by a functional traffic and land-use division. Traffic was separated into pedestrian areas, access streets and a major ring road for through traffic. A centrally located pedestrianized street connected one of the major traffic streets with the older street network (Florentiner Platz and Wolfsschlucht) in the form of large steps (see Fig. 9.2) The main shopping street, the Obere Königsstraße (see Fig. 9.1) was also to be pedestrianized, but this was only fully implemented in 1964, despite the political decision of 1946. The 'step street' was decorated by impressive water basins and fountains and was celebrated in Germany and abroad as a masterpiece of modern town planning.

In other cities some parts of the city centre were newly designed and rebuilt, as in Hanover, Hamburg, Frankfurt and in many Ruhr cities.

There were others which were almost completely reconstructed according to the pre-war plans, like Nuremberg, Munich or Freiburg.

An interesting example is Munich. The first post-war plan of Munich was prepared by the town planner Meitinger in August 1945. It showed that most of the city, especially the historic city centre, was to be rebuilt as before the war. The plan included a few corrections to improve traffic flows for motor vehicles. He suggested arcades along some of the main city centre streets, e.g. the Marienplatz, and the buildings along one side of this square were to be slightly set back. He was in favour of constructing pedestrian passages by removing ruins and buildings inside traditional overcrowded courtyards. These yards should also have shops and beer gardens (which have always been very important for Munich's social life).

In order to keep the parked motor traffic out of the city centre a 50–70m-wide ring road was planned including a ring for parked cars. The ring road, which resembled Stübben's large avenues, included six lanes for carriageways and pavements of 8m width, on each side of which 2m were to be used as a cycle path. Meitinger was of the opinion that everybody who came by car should park outside the city centre and walk. This way he argued no space would be lost inside the city centre, and motor traffic could drive through it. Most of the newly planned buildings should be constructed along the city centre ring road. Meitinger designed in total 4 ring roads and about 9 radial roads, some of which were planned to be of 50m width. Yet the carriageways themselves were small, and only two lanes of motor traffic in each direction were possible. Plenty of space was provided for pedestrians and cyclists, including a wide margin for trees (13.50m on each side).

Adolf Abel had different ideas on how to rebuild Munich's city centre. Abel had made some large-scale pedestrian plans for Wuppertal in 1938 and for Baden-Baden in 1942. He had published an article in which he proposed large-scale pedestrianization in 1942.

After the war he designed a land-use plan for some of the destroyed parts of Wiesbaden (1948), again including substantial pedestrian areas. His ideas of the ideal city were published under the title *Regeneration der Städte* (Regeneration of Cities), in 1950. He demanded the strict separation between pedestrians and motor traffic and elucidated that the separation of transport modes was the only way for the survival of our cities. He used Venice as an example and pointed out that the secret of its success was the physical separation of transport modes (Abel 1950).

In 1946, he was asked to provide a plan for Munich's city centre. The plan was discussed secretly with the local authority in 1947. The main feature of the plan was the strict separation of motor transport from pedestrian facilities. He argued that if an independent pedestrian network could be created then the existing road space would be sufficient for motor traffic. The centre point of his pedestrian town was a newly created square (Marienhof) behind the town hall. From there five pedestrian axes would lead through the courtyards of the town centre which had to be opened. Squares, streets and arcades would complement the courtyards. Shops and housing would, instead of opening to the motor traffic streets, open to the yards, squares etc. The plan was greeted with enthusiasm by the City Council but was soon forgotten, and instead the main urban planner, Leitenstorfer, the predecessor of Meitinger, suggested a more conventional plan which borrowed some features of the Meitinger Plan.

The Leitenstorfer Plan did not solve the traffic problems which became apparent with the increase in motorization. The first General Transport Plan for Munich was agreed in 1963. In this plan some limited thought was given to pedestrianization. The argument used in favour of pedestrianization was to counteract out-migration of the population from the city centre.

Urban road transport policies

In the very early post-war years new traffic and transport investment did not have a high priority. Most important was to repair existing tram and train lines which were the main transport modes, apart from bicycles. Private cars hardly existed.

Hitler's famous car, the Volkswagen was seen by British automobile experts as being 'too ugly, bizarre, noisy and flimsy'; to consider even building this car was regarded commercially as a completely uneconomic enterprise (Nelson 1970). Even so, the 'Beetle' was extremely popular with the military and already in 1945 the Volkswagen factory started to produce cars again. In 1949 the factory produced 46,000 cars annually, comprising 50 per cent of all cars made in West Germany. By 1961, the five millionth Beetle had been produced (Sachs 1984).

The black market had made some Germans rich very quickly and with a new currency cars soon became *the* symbol of the 'rebuilding period' (*Aufbauzeit*). Thus it is not surprising that car use rose very quickly. In 1950 car ownership in West Germany was already 595,000 (2.4 million motor vehicles) (Der Bundesminister für Verkehr 1975), which was about the level of 1933. In comparison, Britain had about four times as many cars in the same year. In 1954 private car ownership in Germany had risen to 1.4 million (4.9 million motor vehicles), and in 1960 there were 4,210,000 cars and a total of 8,004,000 motor vehicles on Germany's roads. It took until 1965 before Germany had about as many cars as Britain (8.6 million). Such growth rates imply some substantial problems in the densely built-up urban areas which had largely not been rebuilt for the motor car. Similar to Britain, the growth rate of motorization had also been completely under-estimated in Germany, and no traffic restraint policy was seriously considered. Criticism was expressed that not enough money was spent on road-building in relation to the growth of motor traffic.

By the middle of the 1950s, road-traffic counts and estimates were started, establishing origin and destination of traffic flows, traffic densities etc. of motor traffic, using different counting methods. Most of these counts did not include pedestrians. Increasingly, techniques and practice from the United States were applied because of the limited experience of German road transport engineers.

In general, the fastest adjustment to motor traffic took place in the cities and towns located in the Ruhr area because of the industrial importance of coal and steel. The 80km-long Ruhrschnellweg, originally built between 1925 and 1930, the most important east–west axis in the Ruhr, was already started in 1954 to be rebuilt to motorway standard. But there were other cities which quickly adapted their street network to the growing demands of motorization.

An example of the new vision of the future town is expressed in the following:

both oppressive and a relief is the thought that the city of tomorrow will mean wide concrete ribbons, multi-laned asphalt distributors, high-rise rectangles of steel, concrete and glass, surrounded by uniform but well-formed housing and factories. Church towers and towers of town halls, monuments and cemeteries will remind us of the city of yesterday surrounded by monotonous interlocked tenement housing and factories. And the city of today? Sixteen years after the end of the war our cities are again without wounds; they appear to be blooming, but they are threatened by chaos. Like locusts eating the fields, so do cars take possession of streets and squares. They nest in parks, open green spaces and in the last little bit of small woodland. They demand in their insatiable greed more and more space. Urban planning, glorified with fantastic designs by idealists of dreamlike simplicity, has to make way for traffic expansion. The human being is pushed to the side: he comes in the double sense of the word under the wheel (Först 1962: 14).

The increase in total road building investment can be seen from Table 9.1. A decline of road investment in actual terms occurred for the first time in 1975 and a substantial decline can be seen from 1980 onwards. Up to 1960 local authorities spent about half of the total road investments on urban roads. From 1960 onwards the majority of road investment was switched to building motorways, Federal and State (Land) roads, possibly reflecting the start of out-migration of the population and with it the need for better inter-urban connections.

Road transport ideologies – the underlying trends

During the 1950s and '60s public transport was still the dominating mode in the large and medium-sized urban areas and it was believed that this

Table 9.1 Net expenditure in current terms on all types of roads in the Federal Republic of Germany (million DM)

Year	Total	Motor-way	Federal roads	State roads	Kreis roads	Local authority roads
1950/ 1951	1 085	57	180	181	127	540
1952/ 1953	1 505	82	201	231	182	809
1954/ 1955	2 033	97	229	350	249	1 108
1959/ 1960	5 107	743	799	666	430	2 351
1964	9 672	1 081	1 926	1 325	921	3 997
1965	9 978	1 061	2 106	1 308	887	4 089
1970	14 892	2 528	2 502	1 907	912	6 290
1973	17 354	4 066	1 853	1 862	876	7 608
1974	17 691	3 927	4 353*		1 040	7 652
1975	17 563	3 915	4 544*		1 101	7 218
1977	18 042	3 525	5 332*		963	7 540
1980	23 086	3 736	6 268*		1 756	10 194
1985	20 367	3 239	5 755*		1 498	8 738
1986	20 950	3 250	5 950*		1 500	9 100
1987	20 700	3 250	6 000*		1 500	8 800

* From 1974 onwards only combined figures of Federal and State roads are available.

Sources: Der Bundesminister für Verkehr 1975 and 1988.

would continue. Thus there was little disagreement about promoting public transport. There was also no serious claim to replace trams by buses, though tramlines were reduced in the coming decades. The trams were still seen to be the best public-transport mode for the future of cities in comparison to diesel buses or trolley buses. Despite that, trams were regarded as a nuisance to motor traffic and in order not to restrict traffic flows it was proposed very early that they should have their own right of way or, even better, be put underground (Hollatz 1954). There were a few who were more in favour of bus use (Schlumms 1961) but it never was a serious proposition. However, most middle-sized towns made this change in later years.

One of the reasons why most transport experts believed in the future of public transport was that the degree of motorization in the United States was evaluated by many as a negative development. There were several ideas on motor transport which were shared by most road engineers, for instance that no buildings should be allowed along fast and major roads. Transport experts saw as their main task to get motor traffic moving and the faster traffic was, the higher was the road capacity. Moving traffic was seen as important for the economic health of the city. The efficiency of junctions was of crucial importance in keeping traffic moving. Roundabouts were regarded as safe but not efficient enough. They were disqualified as being 'brake disks' (*Bremsscheibe*) (Mäcke 1954). The first traffic lights with constant green for motor traffic as calculated speeds were first established in Frankfurt in 1952. They had originally been designed to improve road safety but were later used to increase traffic capacity.

Many road engineers were in favour of what could be called 'total motorization'. They accused the cities of having not enough courage to build roads for the future and having missed the chance to build enough roads after 1945 because they kept the historic city street layouts. Even planners like Hillebrecht saw the main failure and the problem with road traffic as being the fact that the cities were not rebuilt according to the needs of 'modern times', which could be interpreted as according to motor traffic (Hillebrecht 1957).

Most saw the need to separate traffic in order to improve road safety. Some accepted that some road space could be used for pedestrian streets, e.g. Hollatz (1954); Hillebrecht (1957); Korte (1958); Tamms (1961); and Hoffman (1961).

The motto 'Free travel for the free citizen' was accepted by many transport experts and urban planners far into the 1970s and '80s. This ideology has to be understood as the outcome of the belief in the 'free market economy' which was advocated by the Finance Minister, Ludwig Erhard (1949–63) who became Chancellor in 1963. The free market economy was extended to road transport from about 1953 onwards. The journal *Verkehr und Technik* wrote in 1953: 'Constructing roads is not only an issue for the car users; it is an issue for the whole nation.'

There were only a few experts who spoke out clearly against car use in the city centre or were in favour of traffic restraint in residential areas. The number of people who questioned the total acceptance of car use in urban areas increased at the beginning of the 1970s. In the postwar years this was by far the weakest group and there were differences between them in the degree to which they promoted or discouraged car use. The most well known representatives were Ernst May, Hans Bernhard Reichow, Deselaers, Bernoulli and Friedrich Gunkel.

Road safety was the aspect in German road transport policy which appeared not to have received the same priority as in Britain though the accident statistics were horrific and would have been unthinkable in Britain (see Table 9.2 and 9.3). In 1950 2.7 people were killed per 1,000 motor vehicles, the comparative figure for Britain was only 1.1 in the same year. In 1954, 2.4 people were still killed per 1,000 motor vehicles in Germany but only about a third – 0.9 – in Britain in the same year.

Table 9.2 Number of motor vehicles and road casualties in West Germany: 1950–88 (in 1,000s)

Year	Cars	Motor vehicles	Killed	Injured	All casualties
1950	595	2 368	6.3	150.4	156.7
1953	1 126	4 054	11.5	315.2	326.7
1954	1 429	4 868	11.7	317.3	329.0
1956	1 663	5 184	12.3	351.0	363.3
1958	2 812	6 619	11.9	358.0	369.9
1960	4 066	7 797	14.1	438.2	452.3
1962	5 941	9 714	14.5	428.5	443.0
1966	9 577	13 147	16.9	456.8	473.7
1970	12 905	16 783	19.2	531.8	551.0
1974	15 999	20 424	14.6	447.2	461.8
1978	19 633	24 611	14.7	508.6	523.3
1982	22 177	28 158	11.6	476.2	487.8
1986	24 471	31 367	8.9	443.2	452.1
1987	25 264	32 444	8.0	424.6	432.6
1988	26 031	33 505	8.2	448.2	456.4

Sources: Der Bundesminister für Verkehr 1975, 1986, 1988.

Table 9.3 Number of motor vehicles and road casualties in Britain: 1950–88 (in 1,000s)

Year	Cars	Motor vehicles	Killed	Injured	All casualties
1950	2 258	4 409	5.0	196.3	201.3
1954	3 100	5 825	5.0	233.3	238.3
1958	4 549	7 959	6.0	293.8	299.8
1960	5 526	9 439	7.0	340.6	347.6
1962	6 556	10 563	6.7	335.0	341.7
1966	9 513	13 286	8.0	384.5	392.5
1970	11 515	14 950	7.5	355.9	363.4
1975	13 747	17 501	6.4	318.6	325.0
1978	14 069	17 932	6.8	343.0	349.8
1983	15 543	20 217	5.9	328.4	334.3
1985	16 453	21 167	5.2	312.3	317.5
1986	16 981	21 699	5.4	316.1	321.5
1987	17 421	22 152	5.1	306.4	311.5
1988	18 432	23 302	–	–	–

Sources: Plowden 1971; British Road Federation 1980, 1989.

In 1960 the figure in Germany had declined to 1.9, in Britain it was 0.7. Then years later the figure was down to 1.1 in Germany but in Britain it had declined to 0.5. It is only in 1987 that the *gap* between Britain and Germany is closing, 0.25 people per 1,000 motor vehicles were killed in Germany and 0.22 in Britain.

There were many reasons for the high number of accidents in Germany. Not only was the number of cyclists and motor cyclists very high, which always has an effect on the number of accidents, but there was also no speed limit from 1952 onwards, easily passed driving licence regulations, unpredictable roads, with traffic signalling which either did not exist or was not geared to the new demands, and with a careless attitude by many new car owners who may not have understood the danger of car driving.

Before 1952, speed regulations varied with the zone of occupation. In the French occupied zone no speed limits at all were applicable, whereas in the American zone speed limits according to street type and vehicle were established, e.g. in built-up areas no more than 40kph had to be driven (Horadam 1983). From December 1952 all speed limits were dissolved, trucks larger than 2,500kg were not allowed to drive more than 40kph in built-up areas and outside no more than 60kph (Horadam 1983). This situation continued even after the new Road Traffic Regulations of 1956 had been passed. In general, there was a dislike of building pedestrian islands and crossings because they were seen as reducing the speed of the moving traffic. Pedestrians were given priority at 'Zebra' crossings as late as 1964 (Schubert 1966). Before, pedestrians had only been 'safe' at crossings with traffic signal installations. In 1958 a speed limit of 50kph was introduced in built-up areas. There was relatively little research carried out on road-safety issues. Federal research in road safety started as late as 1972 (Formation of the Safety Research Department in the BASt).

Road traffic plans and ideas for the city centre

Studying several city centre plans before and after World War II one can clearly see that all cities made some adjustment to motorization. In most cases, even in the Bavarian cities, streets and junctions were widened, new junctions proposed, avenues of trees cut down and often pavements narrowed. New ring roads or radial roads were designed and existing ring roads enlarged. Many plans left through traffic in the city centre and even the ones which deliberately stated the intention to build ring roads in order to exclude through traffic, for instance Nuremberg and Freiburg, still widened their city centre streets and did nothing to restrict motor traffic. The change of the street layout in the city centre of Dortmund is a good example of the new proposed street widenings in 1950

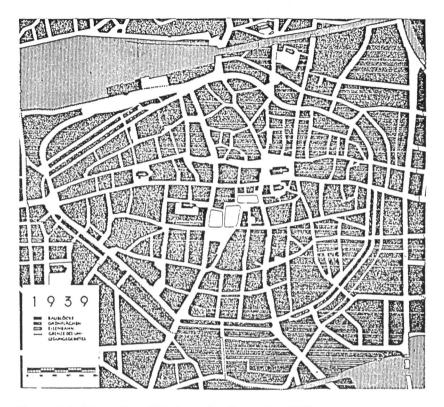

Figure 9.3 Street plan of Dortmund's city centre, 1939.

(see Figs 9.3 and 9.4). In many city centres substantial changes in the street network were made for the first time since the medieval period. This was not only true for the large cities but also for the smaller towns, e.g. Ulm and Heilbronn (DAfSL 1961). The urban heritage was mercilessly made the victim of the car, the symbol of Germany's fast-growing prosperity.

Some ideas expressed then appear extreme in 1989 but they were accepted wisdom in the 1950s and 1960s. There were however a few other representatives who were more moderate. Let us first look at the ideology Kurt Leibbrand was representing. He was Professor at the Technical University in Zürich and had already written a book in 1957 called *Verkehrsingenieurwesen* (Traffic Engineering). He was used by many German cities as a consultant to give advice on how to cope with motor traffic.

Leibbrand's book *Verkehr und Städtebau* (1964) (Transport and Urban Planning) was compulsory reading for every transport engineer and he

Figure 9.4 Street plan of Dortmund's city centre, 1950.

influenced many in favour of his ideas. Separation of transport modes was positively valued but mainly because the faster vehicles could move faster if slower modes were excluded (Leibbrand 1964: 252). Independent pedestrian footpaths were seen as an exaggeration and the upkeep as far too expensive; they were only necessary if motor traffic was really 'disturbing', though no definition was given as to what was meant by disturbing.

Pedestrianization was accepted but pedestrian streets should not be longer than 300m and in general the traffic-free area should not be too large, otherwise the urban area would suffer economically. He pointed out that the word 'pedestrian city' was 'downright absurd' because a town cannot exist without motor traffic. The best retailing locations were the major motor-traffic routes.

He and some other experts seriously believed that car use in the city centre was crucial for its economic health. Concern was expressed that quarters which were difficult to reach by car would lose their value and become slums. Leibbrand was convinced that without motor traffic the death of a city was unavoidable. He gave as an example Regensburg, whose narrow street network could not cope with motor traffic. He wrote:

The narrow lanes of medieval Regensburg are motor traffic hostile. Large parts of the historic centre already show signs of frightening decline. If it is to be saved then more traffic has to enter (Leibbrand 1964: 64).

There were others who were openly against any kind of pedestrianization. Reindl wrote that

pedestrianization was seen by some as the panacea but in reality it implied a bankrupt declaration of existing traffic regulations. Obviously pedestrians will be in favour but one had to consider that streets have not been made for the pedestrian but for wheeled traffic. In the age of motor vehicles it would be anachronistic not to allow cars on the streets. . . . Pedestrian areas have also to be denied because of psychological reasons. They educate people to behave carelessly with traffic and sabotage therefore all the efforts of road safety (Reindl 1961: 3).

However there were a few transport experts, who simply accepted that city centres would never be able to cope with the future demand of motor traffic if one were not willing to sacrifice the city centres totally to motor use. Here are only three examples. Hollatz, Tamms and Korte were particularly important because the first two were later involved in the German equivalent of what Buchanan's 'Traffic in Towns' was for Britain (for more details see below).

Hollatz thought that it was crucial that through traffic was kept out of city centres. It was impossible for him to design city-centre streets which would reduce one-fifth of the city-centre housing area in order to have enough road space. Such estimates were carried out for Hamburg (but similar estimates are known for other major cities, e.g. Hanover and Munich) (Hollatz 1954). He was strongly in favour of large-scale pedestrianization in the city centres.

Korte was in favour of constructing a wide orbital road around the city centre. The additional space needed could come from the old inner city areas (*Gründerzeitviertel*), which were in need of urban renewal. The major radial roads would end at this orbital road. The majority of parking spaces were to be located next to the major city-centre ring road. The city centre was seen by him as an area in which the pedestrian would have priority. Pedestrian areas and quiet shopping streets would be part of it (Korte 1958).

Apart from that, Korte favoured a strict street classification, which was very differentiated with respect to types of residential street, such as residential collectors, residential streets, residential paths and culs-de-sac. Residential streets should only be used for traffic which had its origin and destination there. The major streets should carry the majority of traffic; these should be safe and efficient, the efficiency of traffic junctions was important and had to be built if necessary as flyovers, in order to cope with present and future traffic flows. He used the comparison of the traffic flows as the rhythmic circulation, and streets were seen as branches and twigs. Traffic congestion was regarded as a nervous disturbance of the city which would not be good for its health.

Similar to Korte, Tamms was of the opinion that cars looking for parking spaces should not even be allowed to enter the city centre, and suggested park and ride, and kiss and ride systems, like in the United States (Tamms 1961). He believed that the streets could not indefinitely be improved. Even so, he was in favour of large overall street networks which would include the Federal main roads. But he too saw the establishment of pedestrian areas as important for the city centre.

Transport experts, like Korte, Hollatz, Tamms and even the Deutsche Städtetag (German Federation of Cities) were all in favour of large road-building programmes in urban areas but at least they gave some thought to the pedestrian in the city centre.

If we compare the German attitude with the British one at the same time period then it becomes apparent that the Germans were far more radical in their pro-car ideology but also in the actual road-building investments.

However there was still a small group of German urban planners who had a different opinion on motor traffic which will be discussed below.

Critical views towards motor vehicle use in urban areas

So far, Chapter 9 gives the impression that the majority of urban planners and transport experts were in favour of massive road construction in urban areas. All the past ideas discussed from Chapter 3 onwards seem to have been forgotten. The followers of Sitte, the preservers of the historic heritage and even the planners who had been active during the Nazi era and favoured green axes with separate pedestrian footpaths, appear to have disappeared. That was however not quite correct, though without doubt the advocates of motor traffic had got the upper hand during the first two decades after the war.

However, from 1953 onwards, complaints were expressed about the adverse effect of motor traffic in towns. Traffic congestion and the lack of parking spaces became a serious problem in the densely built-up urban areas. The number of accidents became unacceptable for many. Even so, the majority of road transport engineers believed that relief could be found in one-way streets, construction of bridges and tunnels for pedestrians, investment in public transport, radical street widening but primarily more road building.

There were several urban planners, who warned about the overemphasis on car use. They were for severe restriction of motor vehicles not only in the city centre but also in residential areas. Not surprisingly they were mainly the architects close to the Garden City Society, e.g. Bernoulli, who wrote a short article, called 'Fußgängerstadt' (Pedestrian City) (Bernoulli 1954). The representatives of *Das Neue Bauen* were largely against the unstructured reconstruction of German cities, which was also valid for road building, for instance, Hilberseimer, Häring, Elsässer, Stein. Ironically, many elements and ideas of *Das Neue Bauen* were superficially adopted and reproduced in land-use plans or post-war public building facades.

Hilberseimer had developed a new type of town, which he called *Aststadt* (twigtown) which allowed traffic free access to school and other infrastructure facilities. His ideas remind one strongly of Hermann Jansen's plans (Hilberseimer 1955). However he had very little impact in Germany.

Reichow's 'Organic Town' was similar to the *Aststadt*. His book, called misleadingly *Die Autogerechte Stadt* (The Car Orientated City) advocated for residential areas a street layout which was similar to Radburn. That was indeed not surprising because Reichow had been an active planner during the Nazi time and he had made plans to build a 'Radburn' town in Anklam near the Baltic Lake (today in East Germany) in 1940. After the war be built several housing estates which were closely modelled on Radburn, for instance Hohnerkamp in Hamburg, 1953/54, Steinbüchel in Leverkusen etc. He made plans for an extension of the Margarethenhöhe in Essen (Reichow 1959). His most important settlement was the Sennestadt in Bielefeld which was designed for 24,000 inhabitants and built in 1956.

The post-war road-building euphoria was seen by him only as a large 'surgical operation' which would not solve the general traffic problem. The town as such had to be changed. He wrote that if we do not develop different cities, 'we will become slaves of a transport technology which disdained human dignity' (Reichow 1959: 16). Working and living quarters had to be planned as close together as possible in order to reduce distances.

His main objective was to increase road safety. This was only achieved if the different transport modes were separated. Pedestrianization in the city centre was an important part of his plan. But in the residential areas housing was to be designed in such a way that children should reach schools without crossing roads. Shopping and other infrastructure facilities were arranged on the same principle. He talked about the adverse effects of motor vehicles, the noise and pollution they created and that people had to be protected. Many ideas of modern traffic calming were already included in his book. Pedestrian crossings were to have movable kerbs in order to protect pedestrians from motor traffic.

Like most traffic planners of his time, he wanted to have moving traffic but he pointed out that with his type of town this could be achieved and at the same time road safety could be improved considerably. However Reichow had no impact. His ideas were regarded as regressive.

Ernst May was the most famous representative of traffic restrictions. His housing settlements built after 1945 were similar in their design to Reichow and all had plenty of footpath connections, e.g. Garden City Auemuhl in Hamburg, Osterholz in Bremen.

He made no secret that he thought Germany's transport experts had got it all wrong. In an article published in 1963, he denounced the massive road-building programmes and called them misguided investments and the wrong operations for the organism of a city. It was not the city that had to be changed to accommodate traffic, but traffic that had to be changed to be accommodated by the city. Motor traffic had to be restricted and large parts of the city had to be protected from it. The first step in the right direction was pedestrianization but it was important to extend these areas. He disagreed with parking facilities in the city centre because they would create even more traffic. He suggested instead the promotion and modernization of public transport. He demanded cutting tax advantages for company cars and increasing petrol taxes. The funds raised should be used to promote public transport (May 1963).

There were others who also did not agree with what was going on. Wehner spoke about the city of tomorrow which would be a pedestrian city. The children could grow up without danger to their lives (Wehner 1959). Wehner was later responsible for many transport plans in Germany. Other planners also demanded the modification of car-filled shopping streets into pedestrianized areas (Schubert 1966 and 1967).

Gunkel pointed out that nobody would suggest putting more trains on the rails if it was not safe, to allow more landings of aeroplanes for the same reason, but with motor vehicles we had no restrictions (Gunkel 1965).

Also critical towards motor vehicle use was Deselaers. He questioned the motivation for wide streets in the town centres in order that cars could drive faster than 40kph, and parking spaces should be as close as possible to the shops. He argued that it was not necessary to narrow down pedestrian pavements and cut down the avenues of trees. He accused planning departments of valuing motor vehicle speed more highly than safety, living in peace, the gardens in front of houses and the pleasure of walking (Deselaers 1955). A very materialistic way of thinking had made road traffic a technical-economic objective in itself and the hierarchy of values had been inverted. There was a need to change the residential and shopping streets into living streets by narrowing down the carriageways, and by creating play streets in the residential areas.

At the beginning of the 1960s, there was a kind of helplessness about urban road traffic. Some cities, like Berlin had started to plan massive urban motorways in 1961. (Hoffmann 1961). Other German cities had similar plans. But was that the right way forward for cities, or was promotion of public transport and traffic restraint the better answer? A clear policy orientation was needed on future traffic strategy and as a result of that the Federal Government passed a law which was supposed to answer the most important issues.

Traffic problems of local authorities in the Federal Republic of Germany

The publication called 'The Traffic Problems of Local Authorities in the Federal Republic of Germany' was comparable with the Buchanan Report 'Traffic in Towns' in Britain. This publication was the result of a specific Federal Law (1961) which demanded the setting up of a committee of transport experts with reference to improving the transport conditions in local authorities. The Federal Government had appointed twenty-three experts. Their work started in 1962 and the findings were published in 1965. The group of experts consisted largely of university professors of different transport institutions but it also included representatives of public transport organizations, of an automobile organization (ADAC) and of trade and commerce. Though there was no lack of planners, but only a few were known who had a more critical view towards car use in urban areas.

The main result of the Report was a compromise and was clearly in favour of promoting motor transport. The role of public transport was also emphasized but it was made clear that

it can not be considered to restrain private motor transport; a rational design for urban traffic was needed. A sensible division of the different transport modes was necessary, which implies traffic restraint in particular locations (Hollatz and Tamms 1965: 194).

There were some critical remarks about car use, for instance the warning that 'traffic space for everybody' could not be fulfilled but only four pages were devoted to issues on pedestrians and cyclists. However in many aspects it was written rather vaguely. Professor Gunkel commented on the Report and called it 'tough and leaden reading in contrast to 'Traffic in Towns' which inspires readers with enthusiastic thought. The German Report has a puritanical appearance combined with dryness' (Gunkel 1965: 18).

The Report was very important in as far as it secured the ideological framework for both the promotion of public transport and motor traffic.

There has been speculation on the extent to which the German Report was influenced by the British Report 'Traffic in Towns'. The German publication showed indeed a close resemblance to the Buchanan Report though it is not as clear and logical as the latter, probably because the group of German experts was much larger than the 'Traffic in Towns' team. I am doubtful whether it was strongly influenced by Buchanan, as is often assumed. The German committee was set up in 1961, at the same time as the Buchanan team started, and the first German meeting was taking place in 1962 (Hollatz and Tamms 1965). In August 1964 the German Report was completed (but it was published in 1965) only about ten months after the Buchanan Report. In the German Report, references were made to foreign examples, but none to the Buchanan Report itself, though undoubtedly there were some international connections with and knowledge of the Buchanan Report. Buchanan himself had already outlined his ideas on environmental areas and the role of motorization in several articles and speeches before 1963 and Friedrich Tamms presented a paper at the 'People and Cities' Conference in London, December 1963.

In the German Report, several of the British ideas were missing; notable was the absence of the concept of environmental areas. It included only some vague reference to the Radburn principle (without actually mentioning Radburn). Even more importantly, the issue of restricting motor transport was not sufficiently spelt out. It was clearly not acknowledged as being as important in Germany as the Buchanan team saw it, quite likely because of the more conventional planners who worked on the German Report. Yet much greater importance was given to the improvement of the investment in public transport than in the British equivalent, and it also included a strong commitment to road-building.

Even so, there is no doubt that the Buchanan Report had some influence on German thinking. A German translation of the Buchanan Report was available already in 1964. Strangely enough it was published by the British equivalent of the British Road Federation (Straßenliga). 'Traffic in Towns' (Stadtverkehr) was required reading for every German planner.

Pedestrianization

Despite or because of the large-scale road building investments, pedestrianization in the city centres took place from a very early stage. These early pedestrianization schemes could have been the result of ideas and concepts floated during the 1920s and 1930s. For instance, Kiel had already planned in 1946 to take the motor traffic out of the city centre and to implement pedestrianized areas. The plan also included a wide green belt around the city centre and a pedestrian network which would have connected the green areas outside with the shopping centre (Mäcke 1977). These ideas were combined with an acute problem of road space in most city centres and may have resulted in the introduction of pedestrianization very quickly in the post-war years.

Wilhelmshaven (100,000 inhabitants), Lippstadt (40,000 inhabitants) and Bonn (280,000 inhabitants) were the first cities which closed city-centre streets at least on some days during the week between 1945 and 1948 (Monheim 1975). At that time only Kassel built a modern pedestrian precinct in the city centre. This was very much in contrast with what was happening in Britain; there the pedestrian precinct was the primary form of pedestrianization.

In 1955 21 cities already had traffic-free streets but only 4 had rebuilt the street into a pedestrian layout. As Monheim pointed out, these early traffic-free streets were short, on average about 400–900m (Monheim 1980).

Most cities which closed streets to traffic were located in North-Rhine-Westphalia, and some were neighbouring cities. The reason was primarily that North-Rhine-Westphalia was at that time very prosperous economically and had one of the highest car ownership levels. Traffic separation was accepted by many transport experts and seen as a practical solution. The old tradition of closing streets to motor traffic in cities, like Cologne and Essen must also have played a role. Cities which pedestrianized in these early years were mainly cities which had carried out relatively little street widening in the city centre. None of the Bavarian cities followed the examples of north Germany, and the Ruhr area.

The first south German city to close a street to traffic was Freiburg in 1957 (Monheim 1975). Freiburg had already discussed a plan in 1949 which suggested pedestrianization of the main shopping street (Kaiser-Joseph Straße) but this plan had not been accepted (Vendral 1985). The planning authority decided to close some side streets in 1954 but because of local protest nothing was done. The 1949 idea was proposed again in 1965. This time several small streets were included. Again the proposal was not successful. It took until 1973 to devote large areas of the city centre to pedestrians. The example of Freiburg is symptomatic for many cities. Once the idea of pedestrianization had been discussed, it would crop up again and again, though often in different forms, until it would succeed.

Between 1960 and 1966, the number of cities which had either closed some city-centre streets to traffic or fully pedestrianized streets had increased to 63. Large cities, like Hanover and Düsseldorf started to pedestrianize roads quickly after the introduction of road closures. Before 1972 most cities which pedestrianized opted for small-scale closure and the length of the pedestrian area was mostly under 1,500m; though exceptions were Oldenburg, Göttingen and Essen (Monheim 1980).

Apart from pedestrianization no other policy of traffic restraint was used, except the zone and collar system in Bremen. Bremen was the first German city which successfully restricted through traffic in the city centre. The experiment began before Christmas in 1960. The city centre was divided into four cells. Motor traffic could enter and leave a cell by the city centre ring road. No access was given between these cells (Hall, Hass-Klau 1985). Although the scheme worked very effectively and opposition quickly quietened down, it was not copied by any other major German city. Its first copy was abroad in Gothenburg, Sweden.

By the end of the 1960s, major public transport investment had started and more large-scale pedestrianization was proposed, which will be discussed in Chapter 10. But nobody yet seriously challenged the massive road-building programmes.

Conclusion

After 1945, German cities were largely rebuilt as they had been before the war. Although there were grand new plans in many cities none of them was implemented, largely because nobody dared to change existing property laws and there was a pressing need for housing which made quick action necessary.

Substantial road-building investment started from the middle of the 1950s and continued far into the 1980s. Cars were seen as the most important symbol of post-war economic prosperity. It appears that nearly all transport and most planning experts agreed with the promotion of motorization and cities became substantially orientated towards car use.

The demand to create more and more road space in urban areas for motor vehicles was uncompromisingly followed, not only in the cities, but to a large extent also in the city centres. In the early post-war years most traffic was still planned to run through the city centres. This was surprising because the accepted view even during the 1920s had been that through traffic should be kept out of the city centre.

By the early 1960s it became clear that city centres could not cope with total motorization if large parts of the city centre's housing and shopping were not to be removed. Therefore the only realistic alternative seen at the time was to promote public transport financially by the Länder and the Federal Government. German planners did not seriously discuss any other form of traffic restraint. It is during those years that all major public transport investments were discussed, designed and started.

The Report, 'The Traffic Problems of Local Authorities in the Federal Republic of Germany', was as important as the Buchanan Report. Yet in terms of any general implication motorization would have on urban areas, the Buchanan Report was both more analytical and far-reaching. Also, in contrast to the Buchanan Report, the German Report did not create any controversy over whether it was pro-car or against. The objectives were clear: both the promotion of motorization *and* of public transport were needed to tackle the transport problems of the future. Whereas in the decades before there had been warning voices about too much motor-vehicle traffic in the town centres and in residential areas, and often residential street layouts were designed to allow access traffic only, after the war these warning voices could only be heard sporadically.

An interesting feature during this period was the design of residential street layouts which were similar to Radburn, as developed by Reichow and May and some other architects. The Sennestadt in Bielefeld, Mannheim Vogelsang, Karlsruhe Waldstadt, Frankfurt Nordweststadt, Cologne Böcklemünd and a few more were interesting examples. However such designs were exceptions and not widely copied.

Pedestrianization started quickly in the post-war period but the scale of pedestrianization was small. Although there were many urban planners who wanted more pedestrianization, they were opposed by hard-core traffic engineers who were of the opinion that without motor traffic the traffic-free areas would become slums. There was also the fear of the retailers who believed they would lose out.

Apart from pedestrianization no other traffic restraint policy was suggested. Bremen with its zone and collar scheme was an exception.

Comparing the 1950s and '60s in Germany with the same time period in Britain showed much greater consideration to traffic restraint. Although massive road construction was proposed too, it was mainly justified in terms of relieving residential areas and the city centre from through traffic. Britain had a richness of traffic restraining ideas which Germany totally lacked. Road safety was of far greater importance in Britain than in Germany. German traffic planners also showed a tendency towards extreme views, which was not shared by the British colleagues.

Germany became committed to promoting motor vehicles largely because of the overriding objective to create a prosperous and expansive free economic market which was extended to road transport. The automobile industry established itself very quickly and with an increase in income there was an easily captured home market. The free-market economy was the celebrated way of life and stimulated the demand for consumer goods such as the car, and this new way of life may have been in many cases an 'escape from the past'.

But there were also other explanations. There was a group of transport engineers who had been educated under the Third Reich and became professionally active or continued in the post-war period. This group of experts had very different objectives from the urban planners educated under the same regime. There was also a 'dying out' of the garden-city planners who had their last fling with garden-city ideas during the Third

Reich. As there was little analysis carried out about the Garden City Society itself or its period, anything appearing close to it was confused with Third Reich ideology and rejected. It is interesting to note that both May and Reichow who opposed the 'car mania', belonged to this generation, though politically they were of very different backgrounds.

10 Traffic calming: a new concept for road transport in Germany

The protests against road-building

By the end of the 1960s, critical questions were being raised about the political and social objectives the Federal Government was trying to achieve. Discomfort began to be expressed about the strenuously promoted consumer ideology which many German had seen as the great post-war achievement. This period started with the student rebellion in 1968 and lasted to the middle of the 1970s before the political and social orientation changed more comprehensively. The Christian Democrats had been in power since 1949 and many necessary reforms had not been touched. In 1969, the Social Democrats came into power for the first time with Willy Brandt as the new Chancellor and with the political change great hopes were expressed for a different, more humane society.

Indeed the Social Democrats were more sympathetic towards creating more livable cities and city centres and promoting public transport. However, at the local-government level that was often not the case, and cities like Frankfurt, Hamburg and Berlin were torn apart by immense road construction and office developments which were allowed in the central parts of their cities. In general, there was a small decline of about 14.5 per cent in gross road building investment between 1972 and 1976, but the increase in gross public transport investment was only 8.4 per cent in real terms in the same time period. On average DM 17,000–20,000 million were spent annually on road-building and only about DM 3,000–4,000 million on public-transport investment (at 1980 prices). A marked decline in road construction occurred from 1980 onwards but even by 1987 the relationship between expenditure on roads and public transport was still about 4:1 (not including German Rail).

A change in favour of a public transport policy had already been made in 1967, still under the Christian Democrats, although by the end of 1966 the Social Democrats had already become part of the Government in what was called the 'Great Coalition'.

In 1967 a Federal Act helped for the first time to finance public transport investment. This Act had been the outcome of the Government

Research Report of 1965 which had stressed the need for new public transport investment in order to counteract car use in urban areas (Hollatz and Tamms 1965). The petrol tax was increased by 3 Pfennig per litre and 40 per cent of the proceeds was used to finance public transport investment. In later years the percentage was increased to 50 per cent but in 1989 it was again 40 per cent.

Several publications started to question the role of the car in West German society. Most important was the book by Dollinger with the title *Die totale Autogesellschaft* (The Total Car Society) published in 1972. Though a highly polemic book against car use, it told the story of the overriding importance of cars in German society. Dollinger used among other things a quote from the *Overseas Weekly*, an American forces newspaper, about the characteristics of German car drivers:

When you see the usual driving habits in this country, you must believe all Germans without exception have a death wish complex (Dollinger 1972: 102).

A similar sentiment was expressed by Dahl's book, *Der Anfang vom Ende des Autos* (The Beginning of the End of the Car). This book became very popular because it had been broadcast on radio from 1971 onwards (Dahl 1972).

The student rebellion of 1968 also had some impact on the general attitude towards public transport. It raised the issue of the extent to which cars could be substituted by cheap and efficient public transport. Questions were asked about the main social objectives public transport should have. The major point of dispute was on economic viability versus heavy subsidy for public transport. There was wide support for zero fares for the use of public transport, called the 0-Tarif in order to reduce the number of cars in urban areas. These demands and protests against fares or announced increases of fares, which were also known in several major German cities under the keyword 'Red Point Action' were led by students and school pupils and were politically effective; for instance they achieved the takeover by the City of Hanover of the public-transport system which had been in private hands and had been badly neglected. Red Point Actions consisted of strikes, blockages of public-transport vehicles and gratis taxi-services by students and sympathizers – gratis taxis would have a red point on their window.

It was, among other things, the experience of the late 1960s, which would make a move towards deregulation of public transport as carried out in Britain nearly impossible in Germany.

Although there was a more critical attitude towards the motor vehicle the effects on its use in urban areas was minimal. By the middle of the 1970s, a new overall concept of restraining traffic was discussed by a few planning and transport experts (more details below). Yet the majority of planners were still focusing primarily on promoting new public transport modes in urban areas (light rail, and heavy rail which in urban areas is

Figure 10.1 Munich: pedestrian scheme in the city centre.

called S-Bahn), although research was also carried out on more technically advanced public-transport systems such as mono-rail and magnetic systems (Hass-Klau 1982). This policy was combined with pedestrianization in the city centres. During this time the general 'logic' was to invest in public transport but not to restrain motor traffic.

A new style of pedestrianization

The new style of pedestrianization was demonstrated in Munich in 1972 with the opening of the Olympic Games and as a result it received worldwide attention (see Fig. 10.1). Munich's success may have given many city planning departments the political support to implement larger pedestrianized areas, which would have been difficult to get approved during the 1960s. In the major cities, the precondition of large-scale pedestrianization was a modernized or newly built public-transport system. The combination of public transport and pedestrianization was seen as crucial. Public transport would provide accessibility into the heart of the city. This concept was complemented by large car-parking facilities in close proximity of the city centre and improved roads.

All cities around 500,000 inhabitants and above, had opted for modernized and/or new public transport systems, mostly in the form of light rail, at the beginning of the 1960s. Only two German cities built a new heavy-rail underground network (Munich and Nuremberg) and five urban conurbations had in addition improved and built new suburban rail links (S-Bahn). In 1988, two more S-Bahn systems have been added, including Berlin.

The attention in terms of pedestrianization was focused primarily on the larger cities. Yet the middle-sized towns had also started to pedestrianize on a larger scale but relatively little thought was given to the improvement of the traditional tram lines; that changed by the end of the 1970s. Pedestrianization was also increased and/or implemented in the small and middle-sized towns which relied totally on buses and cars (see Figs 10.2 and 10.3). Many of these towns had scrapped their trams during the 1960s. Here the argument over whether buses should be kept in the pedestrian area or not loomed very early on and no satisfactory agreement has ever been reached, though some towns, like Trier, were very successful in integrating buses into the pedestrian area. The mixing of pedestrians and light rail has never ever been seen as a real safety problem and light rail or trams have been regarded by many planners as part of a pedestrianized area, but often the public transport operators still want to have their trams underground.

From the beginning of the 1970s, research on pedestrianization increased dramatically and the first report on pedestrianization in German towns, called 'Fußgängerbereiche in Deutschen Städten'

Figure 10.2 Bonn Beuel: pedestrianization but access is allowed to trams and servicing vehicles.

Figure 10.3 Buxtehude: pedestrianization and bicycle access.

(Pedestrian Areas in German Cities) was published (Ludmann 1972). It attempted to give an overview on the state of the art in pedestrianization and included directions for planning pedestrian areas.

Without doubt the most well-known German author on the subject of pedestrianization became Rolf Monheim with his publication *Fußgängerbereiche* (Pedestrian Areas) published in 1975. Monheim gave an overview of the characteristics, lengths and classifications of pedestrian schemes in German cities. He also included the different reactions of planning authorities and commercial representatives towards pedestrianization. The book contained street plans of most German towns and cities and showed in the form of maps the growth of pedestrianization over the years. Monheim's publication pointed out that out of the 147 cities and towns which he studied in more detail, about 22 per cent either closed streets to traffic and/or newly designed their pedestrian streets immediately after the lead given by Munich in 1972 and 1973 (Monheim 1975).

Rolf Monheim's later book *Fußgängerbereiche und Fußgängerverkehr in Stadtzentren in der Bundesrepublik Deutschland* (Pedestrian Areas and Pedestrian Traffic in the Town Centres of the Federal Republic of Germany) included a very comprehensive review of literature on pedestrianization and pedestrian studies. In addition it gave an analysis on the historic developments of street planning in several German city centres after World War II. It further contained his own research on methods of pedestrian counts, walking attitudes according to the activities, pedestrian behaviour and use of transport modes by different visitor groups to and in the town centre.

According to Monheim, 370 pedestrian areas existed in 1977. The majority of them were located in the city centres but about seventy were converted traditional local (secondary) shopping streets (Monheim 1986). Since 1977 there has not been a complete survey on the scale and characteristics of pedestrianization. A survey in Bavaria in 1985 showed that out of 151 towns, 82 had pedestrianized or traffic-calmed shopping streets and a further 47 towns were planning to implement such schemes in the near future. In recent years many town centre streets have been traffic-calmed, which allows limited access by cars other than service vehicles but it also includes redesigning the street in favour of pedestrian and bicycle use. There is a trend away from pedestrianization and more towards traffic calming, which will be explained below in detail.

The small towns with less than 50,000 inhabitants in particular have only shown some interest in pedestrianization or traffic calming during the 1980s, which implies that the accessibility to the town centre by car was not seen as a problem for a long time.

Monheim estimated that about 800 pedestrian areas existed, of which about 100 were in traditional local shopping streets in 1986. Most towns with over 50,000 inhabitants have a pedestrianized area. In general, the

Figure 10.4 Traffic-calming scheme: Hameln, town centre.

largest cities have the largest pedestrian areas, some have networks of over 5,000m. But there are also middle-sized towns which have between 2,000m and 4,000m, such as Freiburg, Bonn, Oldenburg, Göttingen, Hameln and many more (see Fig. 10.4).

Pedestrianization as a planning and transport policy became a well-established feature of Germany's city centres. Even representatives of retailing interests were increasingly in favour of pedestrianization, though their idea of a 'good' pedestrian environment was different from the concept of planners or academics. They always stressed the need for sufficient car-parking spaces and good accessibility by car against urban planners who saw pedestrianization as an improvement of the quality of the urban environment which led to people enjoying the newly created urban space. Many traffic engineers also saw pedestrianization simply as an 'exclusion zone' of the town. The majority of the town had to be served as effectively as possible by motor vehicles.

Yet there were some who expressed a critical view about the newly created pedestrianized areas. Dietrich Garbrecht published an article in 1978 with the provocative title 'Fußgängerbereiche – ein Alptraum?' (Pedestrian Areas – a Nightmare?), in which he severely criticized the design standards of existing pedestrian areas, using Hanover as an example. Garbrecht, who has been a keen supporter of the pedestrian, demanded, like a few others, not only pedestrianization areas but large pedestrian networks covering the whole built-up areas. He accused German planning departments of a lack of imagination and monotony in the way they had designed pedestrian areas. He missed the urban excitement which still could be found in many towns in Italy. He coined the phrase 'When you have seen one pedestrian area you have seen them all' (Garbrecht 1978). Garbrecht's criticism hit the spot and an important message became apparent during the late 1970s, which was that the success of (for example) Munich's and Freiburg's pedestrianization was the result of attempts by cities to take into account the specific character of the historic urban structure but also that even large-scale pedestrianization was not changing very much on the dominance of motor vehicle use in the remaining urban areas.

Dietrich Garbrecht published four years later the book *Gehen* (Walking), where he discussed the importance and the different aspects of walking. He pointed out that every third trip and about 50 per cent of all shopping trips were still made on foot (Garbrecht 1981). He described the pedestrian environment which was normally provided in urban areas and which showed that little thought was given to pedestrian facilities. He repeated his demand that not only pedestrian areas should be created, but that even more important was an overall pedestrian network, which should be as comprehensive as the street network for cars. In urban areas there was a need for wider pavements, which should be widened even further in several places, so that benches could be provided and people had plenty of space to stand together, talk and sit down. He wanted more open spaces, soft pavements in parks and more play areas. Residential streets had to become living streets again for all road participants, especially for the pedestrians and the children. He concluded that urban areas would be considerably improved if they were designed in a more pedestrian-friendly manner; thus many car trips could be substituted by walking. It was important that walking in towns became a pleasure. To be in favour of walking, he continued, would result in more 'humanity' in the areas we live in and in the town itself.

'The rediscovery of the pedestrian' and with it the interest in environmentally improved city centres and residential areas was expressed in a number of major national and international publications which flooded the German market by the end of the 1970s, such as Paulhans Peters' book *Die Fußgängerstadt* (The Pedestrian City) (1977) or Klaus Uhlig's book, *Die fußgängerfreundliche Stadt* (The Pedestrian Friendly City) (1979). Books in a similar vein were now also published in English, often including several chapters on German pedestrianization schemes or German planning experiences with pedestrianization, such as Breines and Dean's book *Pedestrian Revolution – Streets without Cars* (1974), Bramilla and Longo's publications *For Pedestrians only* (1977) and *The Rediscovery of the Pedestrian, 12 European Cities* (1974); *International Experiences in Creating Livable Cities* (1981) by Norman Pressman,

Donald Appleyard's book *Livable Streets* (1981), or *Fußgängerbereiche und Gestaltungselement* (Pedestrian Areas and Design Forms) by Boeminghaus (1982).

Attempts at traffic calming

Around the mid '70s when many city centres were pedestrianized on a larger scale, local resident associations of various political colours were formed in many towns to fight against new road proposals and to demand reductions of motor traffic flows in 'their' residential streets. They argued that improvements were needed in the housing areas, in which the problem of increased traffic flows and shortage of parking facilities often became a political issue. In order to fight seriously against new road proposals resident associations had to get advice or even to employ transport experts. The more radical and forward-thinking urban planners and traffic engineers would support such groups and call for a radical change in the existing planning and transport policies. They pointed out that even large-scale pedestrianization and newly built public transport systems had not stopped the deterioration of the urban environment as a whole, and that pedestrianization was only the sweetener for the fully motorized city most traditional transport planners were aiming for.

The formation of resident associations and their ability to become both a nuisance and a serious threat to road proposals has also to be seen in connection with the fact that many voters were disappointed by the Social Democratic Party, which showed little interest in creating more humane cities. Increasingly the publicity on environmental issues, which became known after numerous publications about the possibility of a future ecological crisis, was taken up by newspapers and television, supported resident associations and critical planners in their demand to reduce road-building. Meadows' publication *The Limits to Growth*, a report of the Club of Rome, published in 1972 was immediately translated into German and was widely read (Meadows 1972). The environmental argument to reduce road-building and motor-vehicle use was strengthened by the oil crisis in 1973. Yet none of the established political parties showed any great interest in environmental issues, thus a new party, the Greens, emerged in 1977. The Greens entered nearly all Land Parliaments in the early 1980s and the Federal Parliament in the 1983 election. They gained further support in the 1987 election which has made them the third most important political party. Their political objectives were *inter alia* the promotion of alternative transport modes to car use and a more 'human' urban environment.

Without doubt, the whole issue of finding new ideas for urban transport both influenced and was strengthened indirectly by the political success of the Greens. But the request for a major rethinking in the urban areas was supported too by research and statistics which clarified the way in which substantial socio-economic changes had taken place in the majority of German cities. There was primarily the issue of out-migration of residents, which had started approximately from the middle of the 1960s, and somewhat later, the out-migration of jobs from the densely built-up urban areas not only into the suburbs but, much more worrying for the local authorities, into the hinterland. There was the dissatisfaction of many residents in the newly built social housing estates at the urban fringe. Residents often had to overcome long distances in commuting to their workplaces or travelling to the city centre.

The Städtebauförderungsgesetz (the Federal Law of Urban Renewal) which had come into force in 1971, had changed the character of urban renewal and had given both the local authorities and the residents stronger participation rights and duties. In the past, urban renewal had primarily been seen as improving the housing stock, but increasingly bad housing was also linked with the quality of the urban environment. There was an additional link between a deteriorating urban area and the uncritical promotion of car use. Dieter Apel's research on the adverse effects of motor traffic on the environmental quality of urban streets highlighted that substantial motor traffic was one of the reasons why many residents moved out of housing areas (Apel 1971). His and similar findings were broadly interpreted as 'unfavourable housing and living conditions' by the Federal Ministry of Regional Planning, Housing and Urban Development and included in the Federal Planning Report (Raumordnungsbericht 1972) (BMBau 1973) and again in the Städtebaubericht (Urban Situation Report) in 1974. Because of the involvement of the Federal Ministry of Regional Planning, Housing and the Urban Development (BMBau) in urban renewal, some civil servants of this ministry also increasingly showed an interest in road-transport issues, which was however under the supervision of the Federal Ministry of Transport, and that gave rise to potential conflicts.

It is crucial to understand that for the first time after 1945, the issue of road traffic was again connected with a Federal planning policy such as urban renewal. The improvement of residential streets became part of the whole process of urban renewal and it was increasingly seen as important to create 'livable' streets – *Wohnstrassen* – instead of 'motor vehicle' streets. Such streets would give equal rights to pedestrians, children, cyclists and public-transport users. Over and over again, the adverse effects of car use, specifically with reference to densely built-up urban areas, were stressed and classified into several major nuisances: noise, pollution, high speed, severance and an increased number of accidents involving the motor car and the weaker road users.

Since about 1976, the word '*Verkehrsberuhigung*' 'traffic calming' has become the jargon expression, although there were/are great variations in its meaning. It was largely used for residential streets which were to have less motor traffic or reduced motor-vehicle speed. During the early 1980s transport experts argued about the exact definition of traffic calming but no agreement has ever been reached (Conference on Traffic Calming in Mainz in 1984).

The Federal Ministry of Regional Planning, Housing and Urban Development (BMBau) had already published a report on the role of transport in urban planning in 1974. In the publication *Die Rolle des Verkehrs in Stadtplanung, Stadtentwicklung und städtische Umwelt* (The Role of Traffic in Urban Planning, Urban Development and Urban Environment) a whole range of traffic restraining policies had been suggested (BMBau 1974). About four years later, other reports on this issue were published. Most influential was the Report *Siedlungsstrukturelle Folgen der Einrichtungen verkehrsberuhigter Zonen in Kernbereichen* (Effects of Traffic Calmed Areas – Pedestrianization – on Settlement Structure in Urban Centres), which was a comprehensive study on this topic, including many findings of Rolf Monheim and giving an overview on the economic, social and transport effects of pedestrianization. It pointed out that the loss of population in the city centre and in the areas close by was largely caused by the takeover by other land-use functions, such as retailing and office developments. Large-scale pedestrianization sometimes had the effect of increasing traffic flows in the adjoining residential areas and making them less attractive for living. It advised strongly in favour of overall traffic restraining policies in which pedestrianization only played *one* part (BMBau 1978b).

One year later the publication *Verkehrsberuhigung ein Beitrag zur Stadtentwicklung* (Traffic Calming, a Contribution to Urban Development) was such a success that it sold out very quickly (BMBau 1979). Overnight 'traffic calming' appeared to have become the 'solution' to all the existing road-traffic problems. The importance of the publication lay in the fact that what had been researched and discussed by many planners before now received an official acknowledgement. It repeated the main message which was that urban renewal consisted of more than modernizing and improving the housing stock. The quality of life was dependent on its surroundings, and that included among other things the quality of streets. It continued to point out that urban streets had lost their original character, their social function and their role in providing facilities for the weaker road participants. The main function left for these streets was to cater for motor vehicles.

The Report included articles on several major issues in urban transport, such as the need for a general new orientation of road transport, stronger promotion of public transport, the importance of road safety, examples of new street layouts in different local authorities, and, particularly important, the legal framework for the implementation of traffic calming. The whole range of these issues was now subsumed under the heading of traffic calming.

With these publications the Federal Ministry of Regional Planning, Housing and Urban Development was in effect criticizing some of the past achievements of the Federal Ministry of Transport in terms of road-building.

The claim for traffic calming came from two professional sides. There were the urban planners, the geographers and the architects who had partly implemented pedestrianization or had carried out research on it. They were also often involved in urban renewal. Now they were supported by a handful of traffic engineers who had all worked since the beginning of the 1970s on traffic calming-related subjects. There had been the 'odd' publications on pedestrian issues by traffic engineers, such as 'Planungsmaßnahmen für den Fußgängerverkehr in den Städten' (Planning Measures for Pedestrian in Cities) by Hellmut Schubert as early as 1967 (Schubert 1967). Most important was Konrad Pfundt's publication on *Unfälle mit Fußgängern* (Accidents involving Pedestrians) (Pfundt 1969) or Peter Müller's research on pedestrian traffic in housing areas (Müller 1971). Hartmut Topp worked on pedestrian research in and for Darmstadt in which he showed the importance of pedestrian traffic in residential areas (Topp 1973).

Very important too became a publication by Pfundt and Meewes *Verkehrssicherheit in neuen Wohngebieten* (Road Safety in new Housing Areas). They studied road safety in newly built housing areas and came to the conclusion that the separation of transport modes (pedestrian, cyclist and motor traffic) was crucial to improve road safety. They further concluded that a clear street hierarchy was important, containing short cul-de-sac streets. Motor traffic should be concentrated into major roads and public-transport stops should be accessible by a pedestrian network (Pfundt and Meewes 1975). As a whole, their conclusions were not something outstanding; they had been believed by many traffic planners but for the first time there had been statistical evidence that street design could influence road safety.

From the middle of the 1970s, traffic engineers became more interested in the pedestrian. The German concern about pedestrian safety was also expressed by the opening of a new research department on road safety in the Federal Research Institute (BASt) of the Federal Ministry of Transport in 1972. Several national and international conferences and publications on this topic, such as the OECD conference in Paris in 1975 on *Improved Towns by Less Traffic* (BMBau 1979), the international conference on *Pedestrian Safety* in Haifa in December 1976 and the later OECD conference on *Traffic Safety in Residential Areas* in 1979 (OECD 1979), underlined the international importance of this topic.

Although already at the beginning of the 1970s, a few planning and transport experts had started to demand an overall concept of what was fashionably called 'traffic calming', by the middle of the 1970s this idea became more outspoken and was to strengthen even further in the coming years. The wider concept of traffic calming not only included the pedestrianization of large parts of the city centre but the creation of pedestrian networks, which would connect the residential areas with work and shopping areas, the city centre, recreational and other infrastructure facilities. Traffic-calming measures which implied substantial physical changes to the carriageway of residential streets became the centre of attention.

Crucial too was the promotion of both cycling and public transport. It included a new car-parking concept. There were many planners who were in favour of, for example, higher parking charges in the city centre, reduction of car-parking facilities, not only for long-term parkers, and stricter punishment for car-parking offenders. However in most cases this issue has become the most sensitive subject because existing car-parking policies have been fiercely defended by the retailers, who see any change as precipitating a decline in city-centre trading. Even in 1989, this issue is still not sufficiently tackled in most cities and in some respects rather a lot could be learned from the parking policies practised in Central London.

Traffic calming in residential areas: the practical approach

In the second half of the 1970s, the notion 'traffic calming' was a vague concept. It was a reaction against a vast road-building programme which was destroying traditional urban areas. Traffic calming was not clearly defined and was generally understood by some to include a whole range of alternative urban-transport policies. There were others who simply saw it in terms of new design forms or as engineering measures to reduce motor-vehicle speeds in residential streets. A more theoretical justification for traffic calming came from Britain. The concept of Buchanan's environmental areas was used in academic and political arguments for applying traffic calming. The practical experience of how these livable streets should look came from the Netherlands, where since the end of the 1960s a new street design, called *Woonerven*, had been practised.

The first articles and discussions about the Dutch *Woonerven* and their applicability appeared in the German planning and transport journals by the end of the 1970s, though several transport planners had studied *Woonerven* in the Netherlands before (Kahmann 1979). Delft was the 'magic' place to which German alternative transport planners made pilgrimages. In the next section a summary of the development of the *Woonerven* will be given in order to understand the impact of these Dutch ideas better in Germany.

The *Woonerf* principle – the Dutch approach

In 1963, Niek De Boer, Professor of Urban Planning at the University of Emmen, began the discussion on how to overcome the contradiction between children playing and car use in urban streets by his design suggestions for the new town of Emmen in the Netherlands. His main objective was to design streets, which made some kind of coexistence possible between these two activities. He was of the opinion that street activities by different user groups were a crucial element of an improved urban environment. He designed cul-de-sac streets with additional streets for pedestrians. The culs-de-sac were designed in such a form that motorists felt as if they were driving in their own garden and he gave these streets a different name (*Woonerf*). This design style was very similar to the Radburn street layout, although the culs-de-sac were longer than in Radburn, and was impossible to apply in traditional streets. Those streets had therefore to be changed to force car drivers to take consideration of other road users.

De Boer's idea were taken up by the Municipality of Delft and he was helped by the fact that the road surface in western parts of the Netherlands has to be changed every five years because of the weak subsoil, mostly peat ground, and a prosperous economic climate in the Netherlands around 1965–73 when North Sea gas was discovered and exploited.

At that time, the Planning Department of Delft consisted of four eclectic experts: the principal traffic engineers, Professor Hakkesteegt and Thijs De Jong, the designers and planners A. Kribbe and J. Vahl. Vahl showed that with specific design measures, such as speed humps and trees at the side of pavements, the speed of vehicle traffic could be reduced. They invited the residents who were living in those streets, which were to be redesigned, to participate and express their own ideas. The idea was to avoid the typical street separation between pavements and carriageway. Instead an integration into one road surface was to be provided giving the visual impression of a residential yard. This impression would be enhanced by trees, benches and small front gardens. The planners called these new streets *Woonerven* which is best translated as 'residential yards'. The *Woonerf* idea was implemented in an historic housing area. This area was inhabited by lower-income population groups and students. The students were crucial in the realization of the idea since they shared the enthusiasm of the planners. By campaigning in favour of *Woonerven*, they were able to attract the residents' interest.

At about the same time as the first experiment with *Woonerven* was being carried out in Delft, a new social housing estate, Bÿlmermeer, south of Amsterdam, was built according to the Radburn principle in which the walking distances to car-parking facilities were up to 500m. Both concepts, *Woonerven* and Radburn, were expensive but cost roughly the same. In Bÿlmermeer, among other things a large percentage of the residents was made up of foreigners. This fact contributed to a multiplicity of social problems; vandalism became notorious because car-parking spaces were not directly provided in front of the houses and the cars had to be parked in relatively large centralized car-parking facilities (about 800 cars) and were left without supervision. The *Woonerf* concept in contrast was far more agreeable to the car user because at least people could have their cars right in front of the house and hence were less susceptible to vandalism.

After the success of Delft, the *Woonerf* concept became accepted all over the Netherlands. Most Dutch towns started to implement *Woonerven*. Particularly well-known became Eindhoven and Rjkswyk near Den Haag. By the beginning of the 1970s, the idea had been taken up by a government working group and in 1976, *Woonerven* obtained legal status and a new traffic sign was introduced. The traffic regulation excludes through traffic; vehicles had to drive at horse-walking pace; right of way was from the right-hand side; and car parking was only allowed in clearly marked parking spaces. Pedestrians and residents were allowed to obstruct car drivers at a minimal level.

The construction of *Woonerven* became too expensive after the fall in the gas price in the late 1970s weakened the Dutch economy. Nevertheless, the characteristic ideas were maintained but applied in a simplified manner. The methods consisted of reducing the speed of motor vehicles by specific design measures, e.g. speed humps, bottlenecking, raised junctions etc. The traditional street layout was largely kept intact. Despite the oil crisis the number of municipalities which implemented simplified *Woonerven* grew from 175 to 260 out of a total of 800 local authorities between 1978 and 1980 (van Geuns 1981: 124).

In 1983, the 30kph speed limit was introduced in residential areas. Since then more 30kph speed-limit zones have been implemented than *Woonerven*. Only in specific circumstances, such as the front of schools, will *Woonerf*-type designs be chosen within a 30kph speed-limit zone.

The author is aware that within the constraints of this book it is impossible to portray the whole history of traffic calming in the Netherlands. Important in this connection is that the events in the Netherlands had a strong impact in Germany. Official participation by the Dutch Government (Mr Knappstein), the Study Centre for Traffic (SVT, Mr De Kievit) and the Technical University of Delft (Mr Bach) were made at the discussions about a change in the regulation of the German street design (EAE 85 – Empfehlungen für die Anlage von Erschließungsstraßen) in Wuppertal in the late 1970s and early 1980s.

Figure 10.5 *Woonerf*-type traffic calming scheme, Bonn.

From the German *Woonerven* to area-wide traffic calming: some practical examples

It is rather difficult to trace back where the practical experiments of larger-scale traffic-calming projects started first in Germany. Possibly the first was the research project, which had been launched at the end of 1976 in North-Rhine-Westphalia on traffic calming in residential areas (see Fig. 10.5). The objective was to improve the traffic conditions in local authorities in the form of physical road design (*Woonerven* and related measures) and newly developed traffic signals which, combined together, were supposed to be able to improve the living quality in residential areas. The aim was to:

- improve road safety;
- reduce through traffic;
- achieve slower speeds;
- create more open spaces; and
- provide more trees, shrubs and flowerbeds for the areas.

It was made absolutely clear that traffic calming was only to be carried out in the residential areas and did not affect major roads. The chosen streets would not exceed more than 500 motor vehicles per hour during the peak hours (Der Minister für Wirtschaft, Mittelstand und Verkehr des Landes Nordrhein Westfalen 1979). 'Before and after' studies were carried out. Thirty areas were chosen and thirteen residential areas were intensively redesigned, others partially. The chosen local authorities received funds from the Ministry in North Rhine–Westphalia. Only a short period of twelve months was given to monitor the effects of the implementation of the different design elements.

The criteria chosen in the 'before and after' survey were:

- traffic volume-speed;
- behaviour of motor drivers, pedestrians and children;
- quality of the urban environment which included changes in the safety level (number of accidents, severity of accidents), accessibility, noise, and acceptance by the population.

The measures ranged from culs-de-sac, one-way streets, loops, totally redesigned streets (mixed use), partial change of street character in the form of road humps, raised road junctions (which have about the same level as pavements), bottlenecking, sharp bends in the form of car-parking spaces at a right angle to the right of way, experimental traffic signs, new layout of parking spaces, planting of trees, etc. In most cases a combination of different measures was applied.

The results were extremely encouraging. Most important was the decline in the number of accidents by 20 per cent and a decline in the number of severe accidents by more than 50 per cent. Through traffic was discouraged if a number of combined measures was used, but loops and barriers were not seen as ideal because they increased the length of motor journeys. There was also both a notable decline in the noise level and in the speed of motor traffic though there were differences according to the measures applied.

In 1976 two German planners and one traffic engineer (Max Eiche-nauer, Hans-Henning von Winning and Edgar Streichert) received a contract from a small suburb of Munich where they wanted to apply ideas on a new street layout. After having received the contract they travelled to Delft and were convinced that the Dutch methods could be applied in German towns. They applied for a research grant to the Road Safety Department (Unfallforschung) of the Federal Research Institute of

Road Transport (BASt) in order to carry out research on the effects such *Wohnstraßen* (living streets) would have on a whole range of aspects of road-user behaviour, including accidents and changes of noise and speed level of motor traffic. Although there was scepticism about the methodology at the beginning by the Federal Research Institute of Road Transport (BASt), a research project on traffic calming was agreed in 1977. The case study took place in Unterhaching, in which some of the streets were in need of substantial repairs. These were supposed to be redesigned according to the Dutch examples of mixed use which implied that roads were built without pavements, and some road space was used for trees, lawn and flower beds. Over 9,000sq.m of road surface were rebuilt this way (Eichenauer, von Winning and Streichert 1978).

The major research questions asked were:

- Are the risks for road participants greater when roads are rebuilt to mixed use?
- What effect does mixed use have on motor vehicle noise?
- How do road users behave in these redesigned streets?

The results were surprising, and they were not what sceptical traffic engineers had forecast. The redesigned roads had less through traffic. Far more pedestrians used the streets, they stayed longer and more communication took place. In contrast to what was expected, all road participants felt more secure and their behaviour changed in such a way that they took more notice of each other. Accidents did not, as predicted, increase; they actually declined, though the number of road conflicts stayed the same, but no dangerous behaviour of car drivers was noticed. There was a reduction in average speed which varied according to the change of the road surface, and motor noise had declined because of the lower speeds.

Parallel to Unterhaching another research project was carried out in Berlin Charlottenburg. The local authority redesigned the streets in a small historic tenement housing quarter and at the same time research very similar to Unterhaching and North Rhine–Westphalia was carried out (Keller 1981).

Another important research project was financed by the Federal Ministry of Transport and the motoring club ADAC. They commissioned research on pedestrian safety and the results were used to make recommendations for local authorities. The main part of the report consisted of describing experiences and suggestions on traffic calming in residential streets (Der Bundesminister für Verkehr, Allgemeiner Deutscher Automobil Club 1977). It justified traffic calming on the following grounds, which are still applicable today: traffic calming reduced:

- the number of accidents and their severity;
- average motor speed driven;

- noise and pollution;
- communication deficiencies of residents;
- parking problems;
- rat runs;

A list of ten German towns was given which had already introduced traffic calming.

The results of North Rhine–Westphalia, Unterhaching and Berlin Charlottenburg turned traffic calming overnight into an acceptable transport policy, and 1979 can be regarded as the year in which traffic calming really came into vogue. By 1979 already 96 traffic-calmed areas were recorded, about 67 were implemented in densely built up housing areas built at the turn of the nineteenth century. One of the important learning processes of these early years was that traffic calming in one street would push unwanted traffic into parallel streets. It was thought that the problem could be overcome by an area-wide approach to traffic calming. By 1980 the pressure for traffic calming had become so strong that a new traffic sign (StVO 325/326) was implemented.

In 1980 an area-wide traffic calming project was launched in co-operation with three ministerial research departments, the Research Department of the Federal Ministry of Regional Planning, Housing and Urban Development (BMBau), the Bundesforschungsanstalt für Landeskunde und Raumordnung (BfLR), the Bundesanstalt für Straßenwesen (BASt) and the Federal Research Department of the Environment (UBA). The research combined theoretical analysis in the form of 'before and after' studies with control areas, where no measures were implemented. Substantial financial and practical commitment was asked for from the local authorities but financial help was also given from the Federal Ministry of Regional Planning, Housing and Urban Development (BMBau). After long competition five towns of different size and structure and one village were chosen. The areas in which traffic calming was to be implemented were much larger than in North Rhine–Westphalia. In contrast to the research in North Rhine–Westphalia it included major roads and shopping areas.

The largest city which was to be part of the trial was Berlin; an area of high-density housing built around 1900 (Moabit) was chosen in which 30,000 people live. There was the town of Mainz (190,000 inhabitants) in Rhineland Palatinate, including 15,000 people containing a mixture of an old village and modern housing of 610 hectares. The third city was Esslingen (90,000 inhabitants) in Baden-Württemberg which wanted to introduce traffic calming in an area near the city centre with 11,000 inhabitants. Buxtehude in Lower Saxony with 33,000 population wanted to carry out traffic calming in a large part of the town in which 10,000 inhabitants live (see Figs 10.6 and 10.7). The Bavarian town Ingolstadt which has 91,000 inhabitants, implemented traffic calming in the

Figure 10.6 Traffic-calming scheme: cycle street; Buxtehude: car access is allowed for residents.

Figure 10.7 Traffic calming next to the railway station, Buxtehude.

medieval city centre. There was the small historic village, Borgentreich, of 2,300 inhabitants in North Rhine–Westphalia (BfLR, BASt, UBA 1983). The 'before' research was mostly carried out between 1982 and 1983, though there were some delays in a few towns. The 'after' research was carried out between 1987 and 1988. The south German towns (Mainz, Ingolstadt and Esslingen) delayed the programme, some because of political rows and others because of financial bottlenecks, or both.

The first preliminary results were published in 1985 and 1988 for the three north German towns (BfLR, BASt, UBA 1985; BfLR, BASt, UBA 1988). The results for the south German towns will be published in 1990. Although the published results were not the final summary, it showed for instance in Buxtehude a substantial reduction in the number of severe injuries. The assumed accident costs declined from DM 5.8 million to 3.7 million. As in previous studies the main outcome was that traffic calming had positive effects on reducing noise, pollution and speed levels of motor traffic and it was supported by the population.

The latest survey which covered the general subject 'Transport and Environment' carried out by Difu showed that out of 127 towns, 98 had implemented traffic calming. The Difu report concluded that most of the existing traffic-calming measures were still isolated, covering only a few streets, but in 26 towns area-wide traffic calming had been implemented (Fiebig, Horn *et al.* 1988). A further survey carried out by the Deutsche Städtetag in 1985 also showed the high interest in improving the major urban roads. Sixty-three towns out of 137 participated in the survey and the majority (82 per cent) had already started to plan changes and 73 per cent had made concrete plans for major roads. Apart from increasing road safety, the second most important objective was improvement of the urban quality in these streets (Der Bundesminister für Raumordnung, Städtebau und Wohnungswesen 1988).

The largest project on traffic calming is presently carried out in North Rhine–Westphalia which funds 126 projects in 94 towns and local authorities. About DM 160 million (£51.6 million) are budgeted, of which three quarters are given by the Land Government and the rest has to come from the local authorities. The programme includes area-wide traffic calming in residential areas and the implementation of 30kph limit zones. Traffic calming in major roads has also become more important (for more details see below).

The 30kph speed limit

From the beginning of the 1980s, discussions were raised about the costs these traffic-calmed streets imposed on local authorities and the time period it would take to cover most streets with such measures. Therefore many local authorities abandoned designing *Woonerf*-type streets and

Figure 10.8 Cheapest form of traffic calming: 30 kph speed limit signs in modern housing estate in Freiburg.

experimented with selected measures, which would reduce motor-vehicle traffic speed but not automatically reduce traffic volumes. This form of street design would still come under the newly implemented traffic sign (StVO 325/326) dating from 1980. However, other local authorities had not carried out any changes to the road surface at all but simply put up a new traffic sign which permitted a maximum speed of 30kph in residential areas. Hamburg and Buxtehude led the way in this policy and had started to implement 30kph speed limit signs in 1983. By 1987 already 557 speed limit areas could be counted in Hamburg covering over 1,000km of streets (Morlock 1987). Buxtehude had an area-wide approach of 'tempo limit 30kph' in a large residential area.

Soon many other towns followed the Hamburg and Buxtehude example, often simply because it was cheap to put up a traffic sign and the residents could not accuse the local transport departments of failing to act (see Fig. 10.8). In March 1985 the Federal Government gave the 30kph traffic sign a trial period up to December 1989. In the climate of 1989, it is assumed that this regulation will be made permanent. Many professional discussions focused from 1985 onwards on the issue of whether the 30kph sign alone was effective enough in reducing motor speed. Most transport experts agreed that this was not the case and it had to be supported by physical traffic-calming measures or be strictly controlled by the police. The Deutsche Städtetag carried out a survey of local authorities in 1986, asking whether they had implemented 30kph speed limit. The result was that especially the small local authorities had done so to a large extent but, apart from Hamburg, the larger cities showed considerable reluctance to implement speed limits.

A report written by transport experts of the Ad-hoc-Gruppe 'Geschwindigkeitsdämpfung' of the Forschungsgesellschaft für Straßen- und Verkehrswesen (Research Society of Road and Transport) summarized the most recent German results of 30 research projects in 20 towns and local authorities on the effects of area-wide traffic calming with 30kph speed limit, which included not only the partial rebuilding of roads but also roads where no changes other than speed-limit signs had been made. The following was found:

- traffic calming had reduced the average speed of motor vehicles in these areas, although most drivers do not keep to 30kph;
- there had been a reduction in the number of severe casualties, particularly for the weaker road participants;
- in some areas one could also find a reduction of the total number of accidents; there had been a decline in air pollution and in the noise level; the petrol consumption was lower than with 50kph (Baier, Collin, Keller *et al.* 1989).

These results were encouraging and may force the central government to accept the 30kph speed limit for residential areas, a demand which had already been made by the Deutsche Städtetag in 1988 (Kiepe 1988).

Conflicts and disagreements about traffic calming

Very early on, political differences over the concept of traffic calming developed. One could interpret the argument as being rather similar to the controversy which arose in Britain about the Buchanan Report. In fact, the Buchanan Report was used by both sides as justification for their position. There were the more radical planners and academics who saw traffic calming as an overall concept which included different but related policies. They were in favour of large connecting pedestrian networks and improved residential streets for pedestrians. They demanded far more attractive public transport facilities to be provided by the established public transport operators.

In their opinion, the existing technical achievements of public transport had in many cases reduced the accessibility for the passengers because the traditional public transport network was thinned out. They were strongly in favour of bicycle transport to replace many unnecessary motor trips, but most importantly their traffic-calming concept implied a reduction of the amount of motor-vehicle journeys and a reduction of the amount of motor carriageways provided in urban areas. It also implied that the major traffic roads would be included in the concept of area-wide traffic calming. However it was agreed that different policies had to be used according to the characteristics of the built-up area and the type of street, and that the traffic volume would determine what measure could be applied. That often meant, for example, that in major roads only a reduction in the width and/or number of carriageway lanes, wider pavements, new bicycle paths and trees could be implemented in order to make the traffic situation bearable.

Yet there were others who saw traffic calming as a selective road traffic engineering policy. This included the bunching of through traffic into a few major roads, thus relieving residential quarters from unnecessary motor traffic. They argued that the capacity of the existing road network was very limited, and as a result of traffic calming further roads had to be built or the existing ones had to be widened. Residential use along such major traffic streets had to be changed into other uses. These groups of professionals, largely road engineers, simply reinforced the idea of a street hierarchy, which was not particularly new. The opposition reacted strongly to such suggestions and verbal and written arguments were fought.

However there were many urban planners and traffic engineers, who were caught between both sides and tried to pursue a moderate position.

From 1979 onwards the active campaigning of the more radical traffic calming group had a measure of success. The existing 'RAST-E, Richtlinie für die Anlage von Stadtstraßen, Teil: Erschließung' (Federal Recommendations for the Design of Urban Roads) which had been in practice since 1971, and had been recommended by the Forschungsgesellschaft für das Straßenwesen (Research Society of Roads) was renewed in 1985. The EAE 85 – Empfehlungen für die Anlage von Erschließungsstraßen – (Recommendations for Designing of Residential Urban Streets) had this time been worked out not only by traffic engineers (Federal Ministry of Transport) but also by urban planners of the Federal Ministry of Regional Planning, Housing and Urban Development (BMBau) and others. The work had started in 1979 (BMBau 1985). The new recommendations included all the various elements of traffic calming in different types of streets, such as overall considerations and specific measures on how to modify roads of different traffic volumes.

One of the most essential outcomes of traffic calming has been that it can only work successfully if traffic engineers and planners work closely together. Very quickly a third aspect, namely the 'environmental' engineering, involving the measuring of speed, noise and pollution, became important too.

When traffic calming started, there were no clear regulations or rules on what were the most effective measures to achieve the desired effects. There was a substantial amount of experimentation with different design elements. Assistance was given by several BMBau publications, for instance on design variations (BMBau 1982, BMBau 1985) or costing (BMBau 1983). Most controversial became the BMBau publication *Planungsfibel zur Verkehrsberuhigung* (Planning Design of Traffic Calming) published in 1982. The publication was strongly criticized not only by the Federal Ministry of Transport, which felt that this time the BMBau had really gone overboard, but also by other organizations, mainly because, as they argued, some of the technical aspects of street rebuilding, which had been quite controversial, were stated in the BMBau publication as facts. Even more controversial became *Stadtverkehr im Wandel* (Urban Traffic in Change), which was produced with great style in the form of a journal, understandable by everybody (BMBau 1986). It was to succeed a similar publication, *Wohnstraßen der Zukunft* (Living Streets for the Future) (BMBau 1980), more than a million copies of which had been published in 1980. A storm of protest arose when the 1986 BMBau journal was published but this time several heads in the ministry were rolling and the progressive movement inside the Federal Ministry of Regional Planning, Housing and Urban Development came to a halt.

Apart from the controversy which went on at a ministerial level, conflict also took place in many towns. Without doubt, mistakes were made and often boring standardization of design features can be seen. The first row over street design blew up in Berlin between the historians and the planners. New forms of traffic calming had to be implemented in Berlin, keeping the historic character of the streets. Yet there were others who also argued against the massive rebuilding of streets, for instance Müller and Topp who asked whether traffic calming was a new form of urban destruction? 'Verkehrsberuhigung durch Straßenumbau: Eine neue Art der Stadtzerstörung?' (Traffic Calming: A New Way of Destroying the Urban Environment?). Their main argument was that traffic calming should start in the heads of car drivers and could only be successful if it was fully supported by governments (Federal, Länder and local authorities). Physical traffic calming should be applied only as a supporting measure (Müller and Topp 1986).

The most fundamental attacks on traffic calming were launched by the retailing organizations. They argued that it hindered businesses because driving had become more difficult. Not only would customers stay away but also delivery was more difficult. The BMBau started a counter-attack in researching the 'difficulties' the retailing trade had in traffic-calmed streets and it was concluded that traffic calming had no directly adverse effects on retailing. There were problems but they could be overcome with the aid of cooperation from both sides.

By the end of the 1980s the scepticism of the traditional traffic engineers about traffic calming has still not completely disappeared and many would like not to apply it, but there is tremendous pressure by the population to apply traffic calming.

The overall political climate has become frostier and traffic calming is

Figure 10.9 Traffic-calmed residential street in Brauweiler near Cologne. A resident voluntarily improves the bedding design in front of her house.

Figure 10.10 Traffic calming of a major through road, Langenfeld.

Figure 10.11 Traffic calming of a trunk road, Much.

often seen more as an isolated transport policy and not, as originally intended, as a policy to improve the urban environment and give equal rights to all road users. However there is a new confidence emerging within the population and dissatisfaction is growing about the adverse effects of car uses especially in the denser urban areas (see Fig. 10.9). Thus many local authorities have little option other than to continue their traffic-calming efforts. During the last two years increasingly main roads have become part of the traffic-calming concept and the first trunk roads have successfully been redesigned (see Figs 10.10 and 10.11). In Wuppertal, the first urban motorway in Germany will be traffic-calmed by way of building a tunnel and landscaping the top.

Conclusion

By the end of the 1960s, the achieved values and objectives of the 'free economy' (*freie Marktwirtschaft*) of the Adenauer and Erhard period were being questioned, primarily by the post-war generation. The cities and towns were particularly vulnerable to an apparently unlimited promotion of motor traffic through over-dimensioned road construction, including urban motorways. In particular, in the built-up areas the gap between road-building and car use could not be closed despite tremendous efforts.

The shift of political power from the Christian Democrats to the Social Democrats marked the real end of the post-war period and the attempt to find new, more idealistic values. Included among other ideas was the promise to create more humane cities. The political and social focus in planning was concentrated on improving urban living conditions. The Städtebauförderungsgesetz (Urban Renewal Law) was a political step in this direction. Public transport was seen as important for the future of cities. The 'Leber Plan' intended to move freight transport back to rail but failed because of political opposition. Both the promotion of public transport and the implementation of larger-scale pedestrianization in the town centres was regarded as a major step towards more livable cities. These two policies were seen as the precondition for excluding motor traffic from the city centre. Although city-centre pedestrianization became for many cities the 'jewel in the crown' it did not fundamentally change anything of the transport situation in the rest of the urban areas. The financial effort to built more and more roads was continued.

However more critical land use and transport experts pointed out that pedestrianized areas were only the start of a new transport policy. Pedestrian areas had to have connecting independent footpath links with the neighbouring residential areas and more restrictive measures against motor vehicle usage were necessary in order to improve the quality of urban life. In a figurative sense, there was the demand to extend pedestrianization into the residential areas but it was difficult to find acceptable design forms which were a compromise between the traditional street and pedestrianization. These new demands were primarily expressed in the residential areas of historic tenement housing which were in need of improvement, or in areas which suffered from increased motor traffic as a result of the growth of car ownership, or sometimes even because of pedestrianization elsewhere.

The new idea of traffic calming was a vague concept. Its origin is difficult to trace back but can best be explained by the emancipation of resident associations and local councillors who were not satisfied with the traffic conditions in their areas and saw the destructive effects road-building had in their neighbourhoods. The 'moral' justification was easily on hand when the first critical reports about the damage to the environment were published. The formation of a new political party, the Greens, strengthened the desire to have less traffic in residential streets. Most important was that a 'new' transport concept had to be found to justify a turn in existing transport policies. The Buchanan Report, which had been known to German planners since 1964, was seen as the 'theoretical' basis for a new approach to motor-vehicle use. The practical examples, the *Woonerven*, were noticed in the Netherlands and were seen as a model for Germany too. The first implementations of Dutch *Woonerven* in Germany were carried out in selected towns and cities in North Rhine–Westphalia, the land closest to the Netherlands, but very quickly the *Woonerf*-approach was copied in other towns. Characteristic of these early examples was a new street design which took away the strict division between pavements and carriageways. It changed the overall character of streets and created more space for social activities; it implied giving back again to pedestrians and children the road space they once had.

The streets were made more attractive by planting trees and flower-beds. The design would only work if cars were forced to drive at walking pace or as slowly as possible, something nobody had ever asked car drivers to do before on a public highway. The whole ideological concept of traffic calming or the 'livable' street appealed greatly to many planners because it broadened the field of road-transport planning and gave them a professional foothold in an area that had been hitherto dominated by traffic engineers. Alternative transport modes, such as cycling, became increasingly fashionable with the middle class.

By the late 1970s, traffic calming had become the planning and transport policy everybody seemed to be interested in. Many cities started to experiment with it and the concept changed. There were arguments over whether to keep the pavements or take them out, but it was agreed that traffic calming should not only cover one street but whole residential areas. This new interest was supported by numerous publications from the Federal Ministry of Regional Planning, Housing and Urban Development (BMBau). Most important too became the level of cooperation between this ministry and the Federal Ministry of Transport, a sometimes painful process. The momentum for change was irresistible and in 1980 was symbolized by a new official traffic sign. In the same year a large area-wide project on traffic calming was started, involving three Federal ministries. By then traffic calming was partly understood to include improvement for cyclists and public transport. Many towns started to implement traffic calming schemes but most of them did not cover large areas.

By the beginning of the 1980s, the concept of traffic calming was enriched by the discussion and implementation of a 30kph speed limit in residential areas. Many local authorities were in favour of this policy because it was the cheapest and quickest option. Yet experience showed that by and large in order to reduce speed, traffic-calming measures were necessary.

The discussions and trials on traffic calming which have now been going on for the last fourteen years have considerably modified the conception of motor vehicle use in society. It is now seen more critically than during the 1960s, despite the fact that car ownership is still rising in German and (I would argue) the car is still Germany's best-loved 'pet'. Germany is a long way from condemning the car or even reducing motor-vehicle use on a significant scale; yet the beginning of a 'rethink' can be noticed. The German planners, but also increasingly the population at large, are starting to understand that German cities cannot be built for motor traffic without losing their character and their identity. Maybe one could argue that not very much has been achieved, but fourteen years of rethinking are not long for a mode which has been worshipped for more than sixty years.

Despite the spectacular success of pedestrianization in some towns there are still many towns in which pedestrianization is very small in scale and is still a reminder of the 1960s period. Most important has been that in recent years not only city-centre streets have been pedestrianization but also local shopping streets. In most cases pedestrianization is acknowledged as a policy for avoiding conflict with motor-traffic and counteracting out-of-town and/or new shopping developments; improvement of the urban environment is often regarded as having second priority. But again there are differences according to the town in question. Since the late 1980s, some towns, e.g. Heidelberg, Freiburg, Krefeld, Ludwigsburg and Neuss have started to implement traffic calming (primarily 30kph speed limits on all residential streets) covering the whole urban area. Other car restraining policies and the promotion of public transport and bicycle use are becoming part of the whole concept of traffic calming. West Germany lags in some policies restraining motor vehicles and promoting new ideas on public transport, behind other countries like Switzerland, Denmark and the Netherlands which have an even more comprehensive package of restraining policies for the motor vehicle but the European influence of these very close neighbours will have some impact in pushing the ideas further and as an official of the Federal Ministry of Transport has pointed out, nobody talks about traffic calming, but everybody is doing it.

In the next chapter the developments in Britain over the same time period will be discussed.

11 Protecting pedestrians and residents from motor traffic: the last two decades of road transport planning in Britain

The period of indecision

It was largely during the 1960s that traffic restraint, pedestrianization and further improvements in road safety were discussed in great detail in Britain, but only partly implemented. It seems strange that many of the traffic-calming ideas today had already been established during this period. For instance, Ritter had pointed out that the design of a street system was of decisive importance for road safety (Ritter 1964). Smeed had concluded that there was a clear relationship between the speed driven by motor vehicles and the severity of road accidents (Smeed 1961). Colin Buchanan had warned about the deteriorating effects of motor traffic if society was not willing to restrain the motor car or to provide for it in the form of expensive rebuilding of the urban structure. His concept of environmental areas was in general accepted throughout the country and introduced in numerous town plans (Ministry of Transport 1963). The Ministry of Transport had concluded that road pricing in urban areas was the most effective weapon of traffic restraint (Ministry of Transport 1967b).

In many newly built residential areas and in the majority of new towns adaptations of the Radburn street layouts were implemented.

Most cities and towns had developed comprehensive city or town-centre plans which all included larger-scale pedestrianization. Some of them had both upper-level walkways and streets which were to be closed to vehicle traffic. Both the central and local governments were still committed to completing their planned urban motorways and ring roads.

David Starkie christened the period from the end of the 1960s to possibly the end of the 1970s as the time of 'environmental backlash'

(Starkie 1982). Already during the second half of the 1960s, people started to realize what large insensitive road programmes meant for the urban environment. Residents began protesting against road-building, particularly when personally affected. Protest appeared to have started everywhere, but London was a centre of controversy. The victory of the Labour party in London in taking over the Greater London Council (GLC) in 1973, and with it the cancellation of the urban motorways originally proposed by Abercrombie. has been described by Peter Hall as a classic case of a planning disaster (Hall 1980). Other cities also abandoned their motorway programmes, or at least large parts of them.

As urban road-building became difficult and controversial, the pendulum did not swing the other way to promote public transport, large-scale pedestrianization or any other form of traffic restraint as comprehensively. This was despite the oil crisis in 1973/74 and the cutbacks on road expenditure by Labour in the mid–1970s. However there were some major local authorities and some metropolitan authorities which substantially promoted public transport, for instance South Yorkshire and Nottingham. Funding of new public-transport investment by central government was possible under the 1968 Transport Act, but only a few public transport authorities together with the metropolitan authorities were able to implement new public transport modes, such as the Tyne and Wear Metro. The formation of several environmental groups, such as Friends of the Earth (1970) and Transport 2000 (1972) achieved little in modification of urban transport policy according to their objectives.

The 1970s and 1980s could be described as a period of confusion and revaluation of much of what had been planned and built in the previous decades. Paul Prestwood-Smith has earmarked transport planning during the 1970s as a period of uncertainty (Prestwood-Smith 1979). There are many questions one could ask about this period but answers may be quite difficult to find. For instance, why did planners not, if people did not want roads, implement traffic restraint measures instead? The examples in London or in Nottingham, described below, showed that even when planners were radical, the people affected did not fully accept such changes. A completely new way of thinking was needed for both the planners and the residents, and it was not only the planners who failed.

There was another aspect which did not help. In 1972, local government had been reorganized. Responsibilities on transport issues were not only again divided between transport engineers and land-use planners but also between boroughs/cities and counties. The professions which Colin Buchanan and many others had fought to unite were divided again. What did a society expect if one profession was planning the 'halls and corridors' and the others the 'rooms', a metaphor Colin Buchanan had used during the 1960s? The result could only be disagreement and inactivity. This, combined with a lack of funds from the mid–1970s onwards, could help to explain a period close to political paralysis in decision-making of land-use planning.

Since the 1970s, the conditions for most road users have deteriorated. They have worsened particularly for pedestrians, cyclists and public-transport users through the increase in motorization and the trend to bigger lorries. For instance the number of very heavy (33 to over 38 tonnes) lorries has more than doubled since 1983 (British Road Federation 1989). Furthermore there has been a lack of:

• sufficient protective policies for pedestrians and cyclists;
• construction of bypass roads;
• investment in both urban public transport and rail transport (apart from Intercity links);
• policies to discourage car use in congested urban areas.

The overall image of car driving has been glorified whereas the image of public transport, especially bus use, has deteriorated. New technology has been applied in the car industry but hardly at all in buses. Other alternative modes of transport, like cycling, have not been sufficiently promoted to show a significant increase in usage.

There is little need to describe the overall transport planning policies of the 1970s and 1980s; they have been sufficiently documented by several publications. In the following section an overview will be given of what happened in the two countries in the last two decades in terms of both road-building, the degree of motorization and road safety, because these facts make it easier to understand at least partly why relatively little was done in Britain during this period to protect pedestrians and residents from the adverse effects of motor vehicles and why so much more has happened in Germany, as we have seen in the last chapter.

Road construction and motorization: a comparison between Britain and Germany

A comparison of the road network between the two countries is quite difficult because of the different road classifications. There is however one type of road which can be compared: motorways. The growth of motorways shows clearly the difference in the underlying trends between Britain and West Germany. We have to remember that the size of the two countries is about equal though the spatial distribution of urban areas is different (Germany: 249,000 sq.km, Britain: 230,000 sq.km) and West Germany had only about 4 million more population (Germany: 61 million, Britain 57 million) in 1986. In terms of motorization Germany had 34 million motor vehicles whereas Britain had 23 million in 1988.

The trunk motorway network in Britain increased from 153km in 1960 to 2,880km in 1987. In comparison, Germany had 2,551km of motor-ways in 1960 and 8,618km in 1988. In 27 years Germany built more than twice as many motorway kilometres than Britain. Only the years between 1980 and 1985 show some decline in the growth rate of motorway building in Germany (12.4 per cent) which is nearly identical to the growth rate of British motorway construction during the same time period (12 per cent). But even in 1988 and 1989 more kilometres of motorway are being built annually in Germany than in Britain. In Britain the height of motorway construction was between 1965 and 1975 where the motorway network increased from 557km to 1,932km. This coincided roughly with the boom in German motorway building, though in Germany it continued up to 1980; between 1965 and 1980, 4,088km alone were built. There are no separate statistics on urban motorways in Germany but this network is very small in both countries. In Britain it increased from 13km in 1965 to 94km in 1975 and has grown little after 1977 where it had reached 132km in 1985.

The statistics about the total road network in both countries show the relative decline in road construction in Britain. Whereas Germany had 259,000km of roads in 1960, Britain had 312,436km, about 20.4 per cent more. By 1987 the position was reversed, Germany had increased its road network by 90 per cent to 492,500km whereas Britain had fallen behind and increased its network only by 13 per cent to 352,292km. If one looks at the highway construction expenditure then it becomes clear that even before 1975, Britain spent far less on highway construction than Germany or France. Spending fell further after 1975, but even when the Conservatives came back into power in 1979, road-construction expen-diture rose only slowly and was by 1986/87 only about 6 per cent higher in real terms than in 1978/79.

It would be very important in order to judge the above statistics to have an indication of the quality of roads in the two countries. For instance there are still 76km of two-lane motorways in Germany whereas such motorways hardly exist in Britain. On the other hand a comparison between the overall street network could reveal that very often German main urban roads are much wider and have more traffic lanes than the British counterparts. This may also be the case for German 'trunk' roads.

Between 1970 and 1985, the number of motor vehicles in Britain had continued to rise and had reached 14.9 million in 1970, 19.2 million in 1980 and 23.3 million in 1988. In Germany, the number of motor vehicles had increased even faster. It had reached 16.7 million in 1970 and 35.5 million in 1988. The number of cars in Britain was 18.4 million and in Germany 26 million in 1988. Motor vehicles measured per 1,000 population was 373 in Britain and 486 in Germany in 1987. Clearly some regions in both countries had a much higher degree of motorization.

However, in terms of road safety, Britain has a far better safety record than Germany, though the difference in pedestrian deaths per 100,000 population is not very significant between the two countries; it was 3.7 in Germany and 3.4 in Britain in 1985. Britain's and Germany's pedestrian fatalities per 100,000 population is more than double than in the Netherlands and substantially higher than that of the United States (National Consumer Council 1987: 106).

The number of people killed and injured has declined since 1966 for nearly all transport modes in Britain. Since 1983 a breakthrough has been made in reducing the number of casualties, especially the number of fatalities (see Table 9.3). In Germany, the number of casualties on roads increased up to 1970 and then steadily declined (see Table 9.2). This is valid for fatalities for nearly all modes (except bicycles).

In particular the decline in pedestrian accidents has to be viewed with caution when comparing both countries, because one can safely assume that there has also been a decline in the kilometres walked in Britain and Germany. Even so there may be a difference in the kilometres walked between the two countries.

In terms of road safety the carrying of seat belts was made compulsory in 1976 in Germany and in 1981 in Britain, though the John Adams risk compensation theory claims that this measure has not done very much to improve road safety (Adams 1985). Both countries have stepped up their campaign on road safety but Germany still does not have a speed limit on motorways, though in terms of overall deaths and injuries such a limit would not change the overall road-safety statistics substantially, as only 8.7 per cent of all road deaths and 6.5 per cent of all road injuries were on motorways in 1987 (Der Bundesminister für Verkehr 1988). However, even the smallest reduction would justify a speed limit.

The difference in the speed of road-building, the degree of motorization, the high number of accidents in Germany and surely the influence of historical factors help to explain why Germany had done more in terms of protecting pedestrians and residents from motor traffic than Britain. However this chapter is basically concerned with the events in Britain during the last two decades.

Pedestrianization during the 1970s and 1980s

As has been discussed in Chapter 8, many local authorities started to introduce pedestrianization during the late 1960s. According to my own calculations with data provided by Cashmore, many shopping streets were pedestrianized somewhat later, between 1972 and 1976 but data are only available up to 1981. Cashmore had provided data on 67 examples of pedestrianization schemes: of his sample 13 streets had been pedestrianized between 1967 and 1971; 45 streets between 1972 and 1976 and only 8 streets between 1977 and 1981. It has to be pointed out that

John Roberts has also provided a very large amount of data but his information does not alter the generalization about dates of major road closures (Roberts 1981).

Recently, some local authorities have increased the number of pedestrian schemes, largely in order to counteract out-of-town shopping developments. These new attempts seemed to be more comprehensive than the first pedestrianization wave. This policy has also been supported by a recent local transport note from the Department of Transport called: *Getting the Right Balance*, a guidance on vehicle restriction in pedestrian zones, published in 1988. There, it is pointed out that pedestrian areas are generally welcomed but there is also concern about them, especially by disabled people (Department of Transport 1988). The increase in the establishment of such zones is seen as having a direct relationship with the increase in motorization, and it is assumed that in future more pedestrianization will be carried out. At present there are three Parliamentary Acts relating to pedestrian zones:

- the Town and Country Planning Act 1971;
- the Highway Act 1980;
- the Road Traffic Regulation Act 1984, Section 122.

The Road Traffic Regulation Act of 1984 replaced the Road Traffic Regulation Act of 1967 which for the first time legally allowed the closing of a highway to motor traffic in Britain.

The Town and Country Planning Act of 1971 stated that provision of a pedestrian zone requires planning permission; thus pedestrianization is not only an issue for road engineers but it should also include land-use considerations.

The Highways Act 1980, Section 75 and Section 115A–115K gives plenty of possibilities for traffic restraint measures. For instance Section 75 allows the variation of the relative widths of carriageways and footways. Section 115A–115K allows the improvement of specified highways with trees, benches, flowerbeds etc. Environmental improvements had partly been possible with the Town and Country Planning Act 1971, Section 212 for General Improvement Areas (GIAs) which will be discussed below.

Without doubt pedestrianization has made a leap forward in the last couple of years. The increase in motor traffic, the lack of road space and the slow awareness of environmental issues are finally having some impact. There are many more examples which could have been mentioned here to show this new trend (see Figs 11.1 and 11.2).

Environmental areas and traffic restraint in residential areas

As a result of the Buchanan Report, 'environmental areas' were set up

Figure 11.1 Large-scale pedestrianisation scheme in York.

Figure 11.2 Refurbishment of Sauchiehall Street, Glasgow.

in many local authorities. Most schemes were a combination of some form of traffic restraint and housing improvement. According to Appleyard, about 150 such areas were either planned or in existence by 1973. Nearly all were part of the General Improvement Area Programme (GIA). Between 1969 and 1973 about 900 GIAs were declared. Many of them included environmental improvements, such as pedestrianized streets, children's play areas, trees and landscaping of the area. The major weakness of the GIA approach was the small size of the area, which implied that traffic could easily be pushed into other residential streets.

The best-known and most successful example of an environmental area in London is Pimlico. Pimlico has been described by many different authors. Part of Pimlico was designated as an environmental area between 1965 and 1967 (Colin Buchanan and Partners 1980). It was an attempt to restrict entry of motor vehicles into the area but to allow free movement inside (Appleyard 1981). Colin Buchanan and Partners pointed out that the success of the scheme was largely the result of spare capacity on the peripheral roads whereas Appleyard made the point that some of the traffic had actually disappeared. In general, traffic flows in Pimlico were not very high; the highest were up to 250 per hour. According to Colin Buchanan and Partners, the costs of the scheme were £39,000 and the accident savings £42,000 annually. One should however consider that the cost of accidents estimated by the Department of Transport have been criticized as being too low, and have only recently been increased. The actual savings would have been far higher if environmental improvements, such as the decline in noise and air pollution, could have been valued.

After the success of Pimlico, Barnsbury (Islington Borough) tried to introduce a similar scheme. A Study Team in the London Borough of Islington had published a report about possible environmental improvements in Barnsbury including not only traffic considerations but also suggestions for housing and work places etc. Barnsbury was divided into seven environmental areas, just as Colin Buchanan had advocated. Appleyard has given a detailed account of the political events in Barnsbury (Appleyard 1981). There was a considerable amount of controversy between various residents groups in the area. When the first part of the scheme was implemented in 1970, traffic levels on distributor roads rose between 50 and 100 per cent and traffic flows on internal roads dropped by 25 per cent (Colin Buchanan and Partners 1980). Apparently more residents lost out than gained. Appleyard pointed out that a larger number of people enjoyed lower traffic volumes but also more people lived with higher traffic volumes than before the scheme was implemented. Accidents on the residential roads declined by 64 per cent but rose slightly on the peripheral roads (Appleyard 1981).

A similar scheme was tried in Primrose Hill (Camden Town) but again controversy was significant and no agreement could be reached (Colin Buchanan and Partners 1980). The Borough of Camden had suggested an area-wide traffic restraining scheme but it was abandoned in 1973, again because of disagreement between different resident associations and political groups (Appleyard 1981).

There have been very few other examples of introducing environmental areas described in the literature during the 1970s, but there were possibly many other towns which tried to implement traffic calming and many have had similar experiences to London.

Another famous example which tried relatively early to come to terms with the motor car was the Nottingham zone and collar scheme of 1972. It was carried out in two stages. The first stage was directed at improvement of traffic conditions in the town centre. It was divided into different traffic cells which excluded through traffic.

This policy was first successfully applied in Bremen in 1960. Street parking was reduced by 400 spaces and both bus priority routes and pedestrianization were introduced. These policies resulted in a reduction in the number of accidents by 60 per cent and a reduction of noise level and car pollution.

The second stage was to include higher parking charges in the town centre, and traffic restraint in residential areas with the introduction of cul-de-sac roads and one-way streets. Along the main road corridors priority was to be given to public transport, with bus lanes and priority treatment at traffic lights. This was supported by large park-and-ride schemes. An experiment started west of the town in 1975, including 5 main roads and part of the inner ring road and two urban areas with 11,000 and 44,000 inhabitants (BMBau 1978a). Unfortunately the scheme was not very successful and was abandoned one year later. It failed largely because the public-transport operator increased the fares quite dramatically but also because the population did not accept the restrictive policies.

Traffic calming: the latest attempts

A predecessor of the Dutch version of traffic calming (*Woonerven*) has been called in Britain the 'shared space' street layout. It was first applied in Runcorn in 1966 and Washington New Town. In Telford, another new town, in the residential area Brookside 4, shared space cul-de-sac roads were constructed between 1974 and 1976 (Stevenson 1979). In 1977 the Department of the Environment and Transport published the *Design Bulletin 32: Residential Roads and Footpaths: Layout Considerations*, containing layout guidelines for cul-de-sac roads or short-length access roads in the form of 'shared space'. Many of the 'shared space' roads were combined with landscaping. The 'shared space' approach was criticized by Bowers and Thomson for being potentially dangerous for children, largely because the speed of the motor vehicles was not restricted enough and speeds between 21 and 28kph were recorded. Their advice was that speed in 'shared space' streets should be restricted to 15kph or less (Bowers and Thomson 1984). They also criticized the use of some of the traffic-calming measures, such as landscaping so as deliberately to achieve poor visibility.

One of the most recent attempts at traffic calming has been the Urban Safety Project carried out by the TRRL (Transport and Road Research Laboratory) in collaboration with the Transport Studies Group at the University College London and the Transport Operational Research Group at the University of Newcastle. The project was defined as an area-wide approach and was aimed at reducing the scattered accidents in residential roads by engineering methods. It also wanted to discourage through traffic and 'to help traffic use the main roads more safely' (TRRL 1988). It has been applied in five British towns (Bradford, Bristol, Nelson, Reading and Sheffield). The techniques suggested by the TRRL and others were carried out by local authorities and were similar to those applied in environmental areas in the 1960s and '70s.

The Urban Safety Project started in 1982 and was the result of a number of studies carried out or commissioned by the TRRL. Its main preliminary trial was the Swindon Study, which resulted in a 10–15 per cent reduction in pedestrian accidents.

The project was carried out in the form of a five-year 'before' and a two-year 'after' study, and each residential area which received treatment also had a nearby control area. The number of accidents was about the same in each area (100 annually), and so was the number of households in the study areas (around 12,000). Only Sheffield included more householders but also had the lowest proportion of people with access to cars (36 per cent). The housing characteristics varied considerably from pre–1914 housing to modern housing estates. In Nelson the whole town apart from the town centre was studied. In 1988, the study was completed in Sheffield, Nelson and Reading. The results show a reduction in the number of accidents with the methods applied by the TRRL. The reduction was about 10 per cent, which had been proposed as the objective. The highest savings were achieved in Sheffield (30 per cent) and the lowest in Reading (4 per cent). The accident savings in Nelson and Bradford have not been impressive (Nelson 7 per cent and Bradford 6 per cent, though for Bradford only the first year has been evaluated) (TRRL 1988). When the Reading scheme was implemented in October 1984 there was an increase in the number of accidents; thus it had to be changed after 9 months. The TRRL report stated that the biggest benefit was for the weaker road users, e.g. pedestrians and cyclists, but again the savings were smallest in Nelson and Reading.

During 1987 the Department of Transport became increasingly

interested in the Central European approach of traffic calming (*Verkehrsberuhigung*). In the same year it had published a traffic advisory leaflet – *Measures to Control Traffic for the Benefit of Residents, Pedestrians and Cyclists* – to suggest traffic-calming measures, and was waiting for the response from local authorities which appeared not to have been overwhelming (Department of Transport 1987b). Despite the interest by the Department of Transport in traffic-calming methods, so far it has had relatively little success in developing this initiative.

By the end of 1987 the Department commissioned a report to find out what kind of engineering devices were commonly used by British local authorities to achieve traffic calming (pedestrianization schemes were not included). This report, written by Edward Dalby, was not published by the Department, and it reveals that traffic calming, as practised in West Germany, has only just been started by many local authorities, though no exact dates of implementation were given. The survey was undertaken in February and March 1988, and in total only twelve local authorities were included in more detail, mainly because the grant given to the consultant was very small. Dalby looked at three major changes to the road layout: changes in the horizontal alignment, such as chicanes and road narrowing; changes in the vertical alignment, e.g. changes in the surface texture, ramps, speed tables etc.; measures to give more protection to pedestrians, e.g. shared surfaces, 20mph speed limits and raised surfaces for pedestrian crossings (Dalby 1988).

Most of the local authorities had used road narrowing in order to reduce traffic speed. The second most popular measure was entry treatment (narrowing of street entries). Speed tables, a measure used frequently by German local authorities, was used in Milton Keynes, Barnsley and Rotherham. Even a 20mph zone was tried in Nottingham but not surprisingly with very little success, because of inadequate legal enforcement. Some of the measures appear to have had an experimental feel to them and some looked like straight copies from central European examples, which would be possible after several publications have been made available during the last couple of years but also traffic engineers and planners increasingly travel to Germany, Denmark and The Netherlands. Most of Dalby's report is devoted to showing, for the 31 specific examples, details about speed reductions, changes in traffic flows, safety benefits and costs (on two examples no comments were made).

It is slightly disturbing to find that many local authorities neither measured the effects of speed reduction nor the change in traffic flows when they implemented traffic calming measures. Out of 29 examples, 17 did not measure speed before or after; 10 did not measure changes in traffic flows and 18 pointed out that they had no accident problems in the locations where they had implemented traffic calming. Only 3 examples could prove safety benefits, and 5 local authorities were still evaluating their schemes. Three believed that there was no benefit which could be attributed to traffic-calming measures and 2 had made no evaluation whatsoever. Traffic-calming measures in Bradford (Girlington) which were the result of the TRRL-Urban Safety Project showed both substantial speed and accident reductions (Dalby 1988). In conclusion, it appears that for many local authorities there is no great knowledge of the availability of the different traffic-calming measures and their effect; clearly this constitutes a substantial failure by the Government. The report did not make clear how comprehensively these measures were used, e.g. distances between them, and whether they were applied area-wide or only as isolated examples.

In 1990 traffic calming appears to be the 'flavour of the year' not only for many local authorities but also for politicians and the media. At least discussions on this subject have started in several professional bodies and in many local authorities. The Traffic Regulation Act of 1984 legally gives many options to apply traffic-calming measures. They have been applied sporadically in several boroughs/cities in London, apparently with little upheaval by the residents. In 1988 Pimlico in the City of Westminster appears to be still a model for traffic calming, British-style. Many design features known from German street layouts can be found here. But there are several other towns which have introduced traffic-calming measures, mostly of the *Woonerf*-type (see Fig. 11.3).

Hertfordshire County Council has shown a lively interest in this approach since 1986. In 1988, they committed themselves to a five-year 'town centre enhancement' programme, including nine town centres, but traffic calming is also planned in several residential areas and local centres. One of the first towns which needed improvement was Bunting-ford. The historic High Street had been spoilt by being part of a major trunk road (the A10). Improvements for the town centre were only possible after the bypass was opened in 1987. Though traffic was reduced substantially by the bypass, it was now found that the cars were driving too fast in the High Street because of the straight character of the street and the much lower traffic volumes. Traffic-calming measures were proposed and by the end of 1989 the High Street in Buntingford was redesigned and rebuilt.

Kent is another county which has shown a great interest in traffic calming (Fig. 11.4).

My main criticism on the present use of traffic calming in British local authorities is that it is often understood as a *Woonerf* approach only. It is not seen as an area-wide concept which has also to include major roads, public transport, cycle facilities and other car-restraining policies. Many mistakes the Germans made in the first years of applying traffic-calming measures are now being repeated in Britain. The Ministry of Transport has still not given the go-ahead for financial support for such schemes. However there will be a change in the speed-hump legislation

Figure 11.3 *Woonerf*-type traffic-calming scheme in Peterborough.

Figure 11.4 Traffic calming in Folkestone, Kent.

in the near future in order to give local authorities more freedom to change roads in residential areas.

Other ideas on traffic calming: road pricing (supplementary licensing) and stricter parking control

In the following two sections other road-traffic restraint methods will be discussed, primarily with reference to London. Road pricing and stricter parking control have again become highly topical and controversial in 1989. It could be argued that if road pricing is ever to be applied it will be in London because road-traffic congestion is worse there than in any other metropolitan area. But if London were a success other British metropolitan areas may follow. There is no doubt that the introduction of road pricing would administratively be complicated but the positive effects would outnumber the negative aspects. Much of its success would be dependent on the level of enforcement, the number of vehicles excluded and the price charged.

Implementing effective road pricing (supplementary licensing) in London or any other major city centre would allow several traffic improvement options for pedestrians and residents. It could result in both a reduction in the number of accidents and in air and noise pollution if large-scale pedestrianization and/or other traffic calming policies were implemented, such as area-wide traffic calming in residential roads, *Woonerven* and *Winklenerven* (the latter term refers to traffic-calmed shopping streets involving a mixture between pedestrianization and car access).

Stricter parking control and under particular circumstances the reduction of parking spaces for both the work force and shoppers could also lead to a reduction in the number of motor vehicles in congested town centres. A reduction of parking spaces, particularly for shoppers, can only be successful if an extremely good and attractive public transport system can be provided, which should have its exclusive right of way. As with road pricing, an important issue for car parking is the price charged and the degree of enforcement. It could be argued that similar effects as for road pricing can be achieved in town centres if both the number of car parking spaces (private and public) for firms is restricted and/or very expensive, and the taxation of company cars is changed in favour of the use of public transport (again public transport has to have its own right of way). Park and ride on the outskirts of towns could also have an effect on the number of motor vehicles if car parking spaces in town centres are reduced and if it is combined with other environmental transport policies, e.g. promotion of cycle use, walking. The recent example of park and ride in Oxford is very promising.

Road pricing or supplementary licensing in London

As we have seen in Chapter 8, road pricing was discussed as a serious option even by the Government from about the middle of the 1960s (Ministry of Transport 1964 and 1967). However, as with so many policy discussions of that time it appears to have been a largely academic exercise. Realistic plans for supplementary licensing (which is often confused with road pricing) for private cars in Central London were put forward by the GLC in 1974. The system would have worked in the form of an extra licence fee or a single admission charge. The area controlled would have included the City, most of the City of Westminster, apart from Knightsbridge and Bayswater and the area north of the Marylebone Road, part of Camden, Islington, and Lambeth. The 1974 plan estimated a reduction in the motor traffic flows of one-third in the restricted area. The peripheral roads, such as the Marylebone Road, Euston Road, City Road, Tower Bridge Road, Kennington Lane, and Park Lane would have had an increase of 20 per cent (Colin Buchanan and Partners 1980). There was great opposition to the scheme not only from the Inner London Boroughs which were located outside the area and feared more traffic but also from the public in general. The main criticism was that car drivers had to pay for the use of roads and that it would favour the better-off in several respects. Firstly they could pay more easily, and secondly road pricing would increase the overall travel speed, which again would favour the better-off who could pay and perhaps valued their time more highly. The scheme was rejected by both main political parties but for different reasons.

The second plan, which was put forward by the GLC in 1979 gave more exemptions than the first; it was also cheaper, and was estimated to obtain a reduction of traffic of about a fifth in the control area (TEST 1985). Colin Buchanan and Partners argued that the traffic effects would have been similar to the Supplementary Licensing Scheme of 1974. They mentioned a reduction of vehicle mileage by one-third for the 1979 Area Control Study whereas in the 1974 scheme it was calculated to be 37 per cent (Colin Buchanan and Partners 1980). The scheme would have been in operation between 8.00 and 18.00 and would have worked in a similar way to the 1974 plan. It was calculated that there were 180 entry points but some of them would be closed leaving 70–80 roads for policing by traffic wardens. In order to enforce the scheme, 600–700 more wardens were needed. It was estimated that the financial benefits were to be three times as high as the costs. But the 1979 scheme had the same fate as the 1974 attempt, though apparently the number of opposing responses when it was presented to the public was very low (98) (TEST 1985).

Road pricing or supplementary licensing has been discussed by several authors, e.g. Mayer Hillman, Anne Whalley and Stephen Plowden.

Plowden argued that car travel in London was distorted because 46 per cent of cars terminating their journey in Central London were company cars and 77 per cent have some kind of financial assistance from their employers (Plowden 1987). The high percentage of company cars is certainly important in the consideration of road pricing or licensing in Central London, because if companies were able to pick up a substantial part of the bill, supported even by tax reduction, then road pricing may have little effect.

As congestion in Central London has been increasing, the discussion on introducing road pricing or supplementary licensing has not stopped. Mr Peter Imbert, Metropolitan Police Commissioner, threatened this idea during peak hours by the end of 1987. The draft proposal by the London Planning Advisory Group sees road pricing as the only realistic option (LPAG 1988) for the future in Central London and other London town centres. The latest idea is to charge a daily fee of £5 at peak times in the controlled area. As in the previous schemes, enforcement would be by traffic wardens (M Buchanan 1988). However the Department of Transport is undecided; the Secretary of State for Transport has recently pointed out that road pricing would be a rather unpopular policy and is not being considered. This opinion seems to have softened but it is unlikely that a definite statement will be made in the near future.

The lorry ban and stricter parking control in London

As any form of road pricing was not politically realistic at the end of the 1970s, traffic improvements were sought to be achieved by a heavy lorry ban for vehicles above 7.5 tonnes and stricter parking control. The first proposal for a lorry ban was made in 1981 when a Panel of Inquiry was appointed by the Transport Committee of the GLC. The Road Traffic Act of 1967, Section 6, gave the GLC wide powers to ban lorry traffic. The situation became more dramatic when after May 1983, 32.5–38 tonne trucks were allowed on British roads. A lorry ban was finally imposed in London at night time and weekends in January 1986. Although soon after the dissolution of the GLC in April 1986 opposition arose in some boroughs, a legal court case permitted the continuation of the lorry ban.

More comprehensive parking control in London was established in the late 1950s though it was always seen as a blunt instrument of traffic restraint. By the end of the 1970s, it was generally believed that rising charges and an increased level of enforcement would bring substantial economic benefits. It was estimated that over half a million parking acts were daily taking place in Central London and about two-thirds of them involved some kind of illegal parking.

The outcry for stricter parking regulations led to the introduction of wheel-clamping as an experiment in the City of Westminster and in Kensington/Chelsea in 1983. With the passing of the Road Traffic Regulation Act of 1984 and the Immobilisation of Vehicles Illegally Parked Regulation of 1986, wheel-clamping became legal. In 1988, about 500 cars daily were either wheel-clamped or removed. In London, the areas of wheel-clamping have been increased considerably and included in 1989:

- Westminster, Kensington/Chelsea: 1983;
- The City: October 1986;
- Thamesside Area, Camden: March 1987;
- St John's Wood: April 1987;
- Notting Hill: March 1988; and
- Maida Vale: June 1988.

In addition, in several of these areas cars are not only wheel-clamped but also removed. Scotland Yard assumed that 350,000 offences of illegal parking are daily carried out. Since 1986 the removal of illegally parked cars has been partially privatized. With this new policy 100–140 illegally parked cars can be removed, whereas before only 30–40 cars were being towed away in Central London. In addition, 50,000 parking tickets are issued weekly. These parking offences are expensive – wheel-clamping and removal cost £94 in 1988 – and it is also a time-consuming business. Very recently insurance firms offering cover against being clamped can take responsibility on behalf of any clamped-car owner. According to the Planning Department of Westminster these firms have been quite effective in reducing the threat of wheel-clamping for the better-off car drivers.

Research by the TRRL has shown since 1983, a 30 per cent of decline in illegal parking in Central London. There has been an increase in the number of short-stay parkers and the use of multi-storey car parks.

After wheel-clamping was introduced, the press started a campaign against it. Wheel-clamping became a target of both the right- and left-wing press. In order to operate the policy more effectively and also in order to reduce the onslaught from the press, a large-scale survey was undertaken in 1987: 93,000 citizens of the City of Westminster received questionnaires and about 4 per cent answered; 83 per cent admitted that illegal parking had been a problem for them, and 77 per cent pointed out that since the introduction of wheel-clamping illegal parkers did not use resident parking spaces. In general, the agreement in favour of wheel-clamping was widespread; only 2 per cent were against the action, 28 per cent were in favour of even stronger measures and 39 per cent believed that some parts of the population should be excluded, e.g. residents or retailers. However the main effect of the survey was that the complaints against this parking policy declined dramatically.

Apart from stricter enforcement of parking offences not only in London but in many other British towns and cities, car-parking policy in general is being revised. Local authorities want to take over the task of policing parking which is currently still carried out by the police. Though the income from parking meters and parking spaces goes to the local authorities, they also have to pay the police for enforcing the parking regulations. Traffic wardens are not employed by the local authority but by the police, and in many local authorities there is a shortage of them. Elliott and Bursey gave an example of a district in London more than ten years ago. Kensington and Chelsea had an annual income from meter charges and residential permits of £1,325,000 but £650,000 had to be paid to the police. It was found that 10 per cent of the meters were out of order and 16 per cent of the time was not paid for. Local authorities would gain financially if they could carry out the task themselves because they have a greater interest in increasing the efficiency of parking control than the police. The first local authority which has already taken over this responsibility from the police is the Borough of Camden.

Conclusion

The last two decades have been described as a period of uncertainty and also of environmental backlash. Both descriptions appear to be apt and these sentiments continued well into the middle of the 1980s. Despite that, the 1970s and '80s were clearly impressive in the wide range of ideas proposed for so many different traffic-calming policies. It seems as if nothing was not discussed in this field, and many attempts were made, some more serious than others. All these policies should have, if pursued wholeheartedly, improved the situation for pedestrians and residents but they were merely attempts; the financial support was so limited and the overall situation deteriorated so much more quickly that one can be doubtful about talking of any substantial improvements during this period in most towns and cities. London has been a particularly bad example. The draft publication by the London Planning Advisory Committee came to the conclusion that in the field of car restraint, coordinated land-use and transport planning, there has been almost no effective action for many years.

Talking in general terms this does not imply that in some cases there were no improvements. In recent years there has been a new emphasis on all these policies and it appears that this time they will be far more successful in the implementation of traffic-calming policies than in the previous two decades. Many of them are ideas and plans which have been discussed in Britain for more than twenty years. There has also been a substantial learning process for both citizens and planners. There is a stronger awareness of environmental issues, which may further increase and will help to promote traffic-calming policies. The closer union with Europe may also have some impact on British road transport planning.

The most important characteristics of this period can be summarized as follows:

- As a result of political pressure by residents (but not only because of that) road-building investment declined from the middle of the 1970s and increased only very slowly from 1979 onwards. Though super-ficially this may be claimed as a victory by the environmental lobby, in reality it left and may continue to leave little scope for environmental improvements in some towns. Some of the pressure should have been directed towards environmentally sensitive road-building.
- Implementation of some motor traffic restraining policies were increasingly the result of pressures by the people affected, e.g. residents and retailers.
- During the 1970s and 1980s, the traffic situation has worsened, affecting all road participants, particularly the weakest. Not only has car ownership risen, but also lorries have become heavier and faster. In addition, public transport and the level of footpath maintenance have deteriorated.
- The characteristics of pedestrianization have slowly changed, though only very recently. Whereas during the 1960s and 1970s it was seen largely as a road-traffic policy, increasingly it is applied as part of a retaliation against out-of-town shopping developments. For many it has become transparent that increased road building will not automatically solve the problem of traffic congestion in town centres. In addition, public transport will need substantial investment from central government in order to come to grips with the urban road transport problem. Restricted use for motor traffic is becoming an issue of importance for the economic success of a town.
- There has been very little Central European influence on the overall transport policies in Britain. Britain has in many ways isolated itself from Central Europe. However many British road-traffic restraining policies had some influence in European urban areas. Road pricing is being seriously considered in Norway, The Netherlands and Sweden.
- The wide range of options for traffic calming have hardly been taken seriously by many transport planners. Since the middle of the 1980s, there are signs that this is changing.
- Traffic-calming measures, such as those used in the Urban Road Safety Project by the TRRL, have been seen purely as a traffic engineering approach and not as part of improving the urban environment.
- Many of these shortcomings can be blamed on the central governments, both right and left, which have underestimated the adverse effects of motor traffic and the potential of traffic calming for the present and future development of cities.

12 Protecting pedestrians and residents from the effects of wheeled and motor traffic

Several principal questions were asked at the start of this book. One was whether road transport policies today are solely concerned with engineering skills, or at best, a mixture of engineering and urban planning, or whether they have been additionally moulded over the years in a confused way by socio-economic and political processes. I have shown that these historic considerations have indeed played an important role and that they present difficulties when it comes to shifting transport policy into new directions.

Misjudgements of one or several variables in transport planning and in our case in the field of road transport, such as the extent of motorization, incorrect emphasis on road building, under-promotion of alternative transport modes, are paid for by society often many decades later. Such mistakes or non-decisions are sometimes seen as being made merely by a previous government; but in reality they could have been made more than a century ago. As a result of not knowing when and what were the misjudgements of the past, they are easily continued in the present and future. There is too little knowledge of historical developments in transport policy and transport planning and if there is, it is often misinterpreted or incomplete. Maybe in isolated instances it helps to avoid political embarrassment. For instance, some of the ideas of the alternative German political party, the Greens, can also be found in the Nazi ideology and the German garden city movement.

The overall theme of this book has been how two societies – British and German – reacted and coped with one of the most powerful technical inventions of the late nineteenth century – the motor vehicle. Although one chapter dealt with street planning in the United States, this was largely used to explain events and influences in German and Britain (and vice versa). The success of the motor vehicle has been told from many different angles and it includes the history of urban street planning, traffic engineering, the struggle to improve road safety, urban planning ideologies, and the socio-economic history of the countries involved.

The leading actors have been architects, engineers, planners and politicians. All of them were involved, and often at loggerheads.

Dominant in the theme has been the relationship between the need for pedestrians to be protected from wheeled and later from motor vehicles and the moral obligations of the two societies in attempting to fulfil this task. Yet it was not only the pedestrian who was at stake, it was also the urban fabric which was largely not designed to cope with motor vehicles and was therefore exposed to fundamental structural and socio-economic changes.

During the last century the main policies for protecting pedestrians and residents have not changed very much. They have included a variety of methods of separating the pedestrian from wheeled traffic, which were used simultaneously, sometimes for different parts of the urban environment. Likewise we can also find in the residential areas a policy of mixing traffic modes.

From the middle of the nineteenth century urban streets were provided with pavements in order to separate pedestrians from vehicles. More protective inventions were arcades or road closures. In residential areas the newly invented street design, such as the narrow and short culs-de-sac or the independent pedestrian footpaths, had a similar function.

Another main question was to enquire into the fundamental differences and similarities between transport policy and planning primarily in Germany and Britain. If there have been variations, how can they be explained and do they give us the key to modern transport policies with reference to protecting pedestrians from wheeled traffic? Without doubt both countries are entrenched within a complex mixture of history and attitudes, which are nearly impossible to disentangle, but there are patterns which tend to be specific to one country only. They appear at the beginning of the book and can be found repeatedly throughout. One element has been the different attitude towards the city as such by decision-makers, which appears to be still present during the 1980s.

By the end of the nineteenth century, Britain and Germany developed their first ideas about town planning. It is rather interesting to note the difference between the two countries about what was the 'model' of an ideal human settlement. In Britain the preferred form was the village; the reader may remember the beginning of Parker and Unwin's first book which started 'as beautiful as an old village'. The romanticism of village life has continued up to the present day.

In contrast, the majority of Germans never had this exaggerated love for rural life. It was generally seen as being backward, boring and uninteresting. In Germany it was the medieval, and to a lesser extent the Renaissance town which was the perfect place to live in. The reasons for the different idealized urban settlement forms in town planning cannot be fully explained here, and may be regarded as purely speculative, but interpretations could lie in the fact that Germany had lacked a political unity for centuries and although the Germans desperately longed for one, the only substitute they had, before unification in 1871, was the town and region. Possibly more important was the mixture of romanticism and patriotism which glorified the medieval period itself – a time of expansive political German history. Primarily from the middle of the nineteenth century, the medieval urban structure was regarded as worth preserving under most circumstances. The whole medieval era was something of a sacrament for the Germans and it is no accident that one of the Nazis' first actions was to announce the careful restoration of the medieval city centres which had suffered badly under previous governments because of lack of funds.

Industrialization arrived later in Germany than in Britain and also affected the German cities differently. By and large industries settled in existing and often rather important political and administrative towns, though exceptions can be found in the coal regions, such as the Ruhr, Silesia and the Saar.

In Britain, many important industrial towns had emerged from mining settlements, villages or small harbours, often with little historic urban inheritance. It is beyond the scope of this book to analyse the fundamental differences in industrialization but it is clear that the speed of industrial growth and the location of factories and working-class housing in cities were very different in Germany and Britain. Most of the German city centres and many of the residential areas close by remained, even at the height of industrialization, very livable places for the rich and the upper middle classes, although socio-economic changes did occur in the city centres and neighbouring residential areas. But these changes were not as rapid and drastic as in Britain.

Peter Hall in his recent book *Cities of Tomorrow* has revived the phrase 'City of Dreadful Night' describing the large nineteenth-century slums. This phrase embraces strikingly what the slums all over the world were like. But industrial cities did not consist only of slums, and maybe many Germans did not want to have their cities totally identified with this grim aspect of urban life. Yet in Britain there seems to have been a consensus that cities were doomed, and it was better to get out or at least live on the outskirts. The Germans had their own ideologists who 'witch-hunted' city life but there was always a counter-balance, the ideology-forming strata of society who found city life exciting.

That is not to say that British cities did not have civic pride; many of the nineteenth-century buildings and squares in the major cities show that very vividly. But civic pride in Britain appears not to have been as deeply rooted as in Germany. Many British city-dwellers may not have been able to identify themselves with these newly created urban 'spaces' and saw them only as symbols of wealth and political power. In Germany the situation was very different, and even most of the newly constructed buildings were replicas of Germany's glorious medieval past. In addition, they had their narrow, crooked streets, their historic market squares, the densely built medieval houses and their cathedrals. Surely there are plenty of medieval cities in Britain too, but during the early period of industrial development few of them could attract industries and become wealthy, as many of the German medieval cities did.

The German attachment to the historic city remained strong through the decades. Even the German garden city movement created only small garden suburbs of no more than at most 2,000 dwellings. The majority of them had close 'ideological' and functional connections to the existing city. In the coming decades, the Germans always kept these links to the 'mother' town. Even Ernst May's concept of the satellite towns can hardly be seen as an exception. This idea was evident too in all the other big housing estates built during the Weimar Republic and did not really change after World War II. Thus Germany has no new towns in the British sense. The two exceptions, Wolfsburg and Salzgitter, built during the Nazi period were not explicitly planned and were the result of unusual circumstances. In contrast, the objective of the British garden city, and after 1945 the new town movement, was to create independent towns of considerable size.

The ideological importance of the German city had strong influences on the attempts to preserve the city centre and adjacent historic parts very early on from influences which would change its structure. Clearly the need to adapt cities to the motor vehicle demanded some quite substantial changes in the urban structure and this had to be opposed. Let us in the next section discuss the variations in city centres in both countries in more detail.

Traffic restrictions in the city centre

Germany

In Germany, Baroque street and square designs were built from the seventeenth to the nineteenth century to create among other things improved conditions for wheeled traffic. Despite the great power of the German sovereigns, they respected the medieval town centres in nearly all cases. Even the most comprehensive street plan in Berlin, the Hobrecht Plan, did not touch the medieval city which was tiny and unimportant in comparison to cities like Cologne, Frankfurt or Nuremberg. This has to be seen in contrast to Paris where the Haussmann plan destroyed large parts of the medieval street layout.

In general, medieval street patterns have tended to protect the pedestrian from wheeled traffic simply because many streets were too narrow for the passage of wheeled vehicles. Indeed, many city-centre streets presented serious conflicts, sometimes exacerbated by the space surrendered to tramway systems. There were several options to improve the traffic situation. One was to put forward 'modern' street plans as advocated by Reinhard Baumeister, another was to cope with the existing structure by closing streets, introducing one-way circulations and by diverting as much traffic as possible out of the city centre. The easiest way out was a compromise, such as the removal of the medieval walls, during the second half of the nineteenth century, which was carried out frequently in order to get a wide ring road as an alternative to the expensive redesigning of city-centre streets.

It seems that road closures were common from the middle of the nineteenth century onwards. Unfortunately, the police orders of the time do not tell us the reasons for such closures and we can only assume that it was to avoid conflicts and to protect the pedestrians. We know that the local police were very powerful and that there was a very close connection between the issue of traffic restraint and the 'preservation of monuments and historic urban structure', from 1900 onwards.

Despite the attempt to keep the medieval street pattern, we know of numerous cases of street widenings or newly built streets in the city centre, but such actions were strongly disapproved of by Sitte, Henrici, Rebhorst, Gurlitt, and many more. Yet, by and large, many German city centres kept their old street structure until the 1950s. Hamburg was a great exception and was branded as 'Free and Demolition City Hamburg' (Freie und Abrißstadt Hamburg).

During the Weimar Republic road closures continued. But now, there was an increasing number of planners and politicians who argued that the city centre had to be changed in order to accommodate the growth of wheeled and motor traffic. That not very much was carried out was largely the result of economic circumstances and the more pressing need for social housing, but also the opposition which still existed towards demolition of valuable historic buildings and squares in order to get wider streets. The Nazi period was a continuation of the discussions which had taken place during the 1920s. Attention was focused on motorway building and thus road-building in the city centre had lower priority and was often seen as not necessary any more. We still find evidence of road closures. Pedestrianization was now sometimes deliberately planned in Hitler's newly designed towns in the forms of squares or wide avenues in order to honour the Führer, but apart from that there were already several serious discussions of town centre pedestrianization.

After World War II, publications like Adolf Abel's on creating a pedestrian city in Munich were resurrecting ideas of the 1930s and '40s. The number of pedestrianization schemes increased during the 1960s; indeed by now it had become a deliberate policy. However it was questioned by experts like Leibbrand who announced that city-centre streets would become slums if they were pedestrianized. Perhaps this was because motorization was seen then as a symbol of progress and 'freedom' of the mobility of the citizens.

The demands of the 1970s for large-scale pedestrianization, pedestrian networks and traffic calming of the residential areas was the fulfilment of demands first expressed at the end of the nineteenth century to reduce the impact of motor traffic, to protect the urban heritage, and to create 'livable' cities.

Today the most remarkable German pedestrianization schemes are connected with towns which have a medieval core, and I believe this structure has helped to push pedestrianization to the success it has received world-wide. It is interesting to note that West Berlin (the medieval city centre is in the East), and Hamburg, which had already changed its street layout substantially by the end of the nineteenth century, do not have a large pedestrian network. But it would be wrong to overemphasize the aspect of the medieval street layout; surely the deep-rooted German commitment to urban inheritance, its history of road closures and the protest against massive urban road building programmes after World War II, are much stronger elements to explain the development of large-scale pedestrianization in Germany.

Britain

If we compare the German city-centre structure with the British, then we find that most of the large cities in Britain did not have an intact medieval street network. Plans for new street layouts common on the Continent during the nineteenth century were heatedly discussed but put into practice only in fragments. Urban residential streets, built during the same period, were regulated by bye-laws which created relatively wide streets, often with frequent intersections. Apart from arcades and traffic management which already included pedestrian islands and traffic lights, evidence of road closures is difficult to find. Although some cities like London were plagued by traffic congestion and conflict, the government policy going back to the Street Act of 1867 stressed the need to keep streets open for all road participants.

The Manchester Corporation Act of 1934 allowed non-classified streets to be converted into play streets, and similar Acts existed in other towns but this was a policy primarily concerned with residential areas.

During the 1930s, precincts, arcades or upper-level walkways were discussed in order to protect pedestrians from motor traffic. Similar suggestions, apart from precincts, were known in Germany and sometimes open arcades were built in the town centres.

After World War II, pedestrianization was carried out, not in the form of road closures but largely as purpose-built precincts, which we find in

254 The pedestrian and city traffic

several of the new towns and cities rebuilt after war damage. In that respect the British approach was closer to the Dutch idea of Lijbaan in Den Haag which opened in 1953. One could also argue that because an identifiable historic city centre was often missing there was a continued attempt by city planners to design and define a city centre according to their own ideas which may have had little to do with what the citizens actually wanted.

It was the Buchanan Report which had a substantial impact on the further advance of pedestrianization. Buchanan freely admits that he visited other European countries in order to get new ideas for Britain. Yet again, though Buchanan suggested pedestrianization 'German style', he was also in favour of upper-level walkways. When finally, in 1967, street regulations were changed to allow the closure of highways, and pedestrianization of streets took place, even then the purpose-built shopping centre became the focus for pedestrianization in many British cities.

When looking at the history of pedestrianization in Britain, we notice that the British plans to protect pedestrians were far more ambitious and more expensive than the German way of closing roads, and therefore more prone to failure.

It is often argued in Britain that purpose-built shopping centres are more common because the weather is so unpredictable. I do not think this argument stands up to examination. One of the reasons why British pedestrianization is so different to Germany is in my opinion rooted in another 'city culture', which I have tried to explain above, but there were other factors of importance. There appears to have been (and still is) a greater acceptance of wheeled, and later motor, traffic as a way of life from very early on and a possible fear of conflicts if 'equal' rights of all participants were not provided. One could argue that Britain was a more democratic country than Germany and respected the rights of every mode without realizing that the weaker road participants needed more protection.

The different emphasis in the public transport system may have also played a decisive role. The extraordinarily dense tram network in all German cities (especially in the city centres), and the far lower degree of car ownership and bus operation before World War II, meant that a larger proportion of the population than in Britain would either walk or use trams, and therefore the closure of roads to motor vehicles was politically more defensible.

Even today, some of the most successful British pedestrianization schemes appear to be in smaller and middle-sized towns which often do retain a medieval street character. Yet even there, we hardly ever find the large-scale pedestrianization schemes known in Germany. Turning now to traffic separation and traffic calming in residential areas the differences between the two countries are not so substantial.

Traffic calming in residential areas

Britain

Ideas for new kinds of street-layouts were developed by the designers of the English industrial model villages and the garden cities. Raymond Unwin and Barry Parker were the most outstanding names associated with that era. For the first time during the nineteenth and early twentieth centuries, streets were designed according to their traffic and urban functions with much concern for the safety of pedestrians. Such streets could also be justified in terms of reduced costs. The design of cul-de-sac roads allowed access-only traffic, and narrow streets reduced in later years the speed of motor vehicles. Varying street widths were used in the British garden cities and garden suburbs and the residential street network became more sophisticated over time. Many residential areas built during the coming decades in Britain continued with this street design, although narrow streets with wide green margins and footpaths often disappeared during the 1960s in order to widen the streets.

But the most important and the earliest concept of 'traffic calming' in a modern sense was developed by Alker Tripp. His concept of precincts which would have had the character of excluding through traffic from residential areas and city centres already suggested vertical and horizontal changes of the alignment in order to make residential streets unattractive for through traffic. But Tripp was also in favour of massive road building. His concept was used in nearly all the major new-town plans during the early 1940s.

After World War II Britain applied a residential street layout which showed a strong similarity to Radburn in the new housing areas but it is not clear to what extent it was a copy of Radburn or a continuation of British street layouts developed during the 1920s and '30s.

By the middle of the 1960s, Britain started to implement environmental areas after the Buchanan Report was published. Several of the main ideas had been expressed before and went partly back to the British garden city movement. Even so, Colin Buchanan had for the first time pointed out the whole spectrum of adverse effects motorization had and would have on the urban environment. However the tragedy was that Buchanan's environmental areas were introduced during a time when society was playing down those effects. Few people understood the complexity of his implications and it was misinterpreted as a charter for massive road-building. It is interesting to note that the Germans published a similar report only one year later which also had little impact on changing the priorities of transport modes.

Germany

In Germany the British garden city movement was eagerly adopted by the Germany Garden City Society. The Germans copied many of the British house and garden designs but few characteristics of the street layouts; the cul-de-sac was one exception but it was not used frequently. In fact, street planning was already well advanced compared to Britain. Variations of street widths were common and the importance of the different traffic functions of streets were stressed in the publications of Reinhard Baumeister, Hermann-Joseph Stübben and Camillo Sitte at the end of the nineteenth century. Sitte demanded the protection of the historic urban structure from the onslaught of wheeled traffic. His ideal street was narrow and curved as it had been during the medieval period. It was largely Sitte who influenced the street design of the German garden suburbs although not all were provided with crooked streets.

In these suburbs, and also in the contemporary housing areas, great importance was given to communal greens, even when private gardens were planned. These were connected by narrow streets, which were sometimes intended to be used as footpaths only. The residential street network had three or four standards of street width. In fact, these early street designs came very close to traffic separation between the pedestrians and wheeled traffic, and it could be argued that they were the predecessors of the Radburn design, built in the United States towards the end of the 1920s.

It appears strange that the Germans were so obsessed with the idea of narrow streets when there were so few motor vehicles around. As has been argued above, the model of the ideal settlement was the medieval town. Thus the main characteristics of garden suburbs and similar housing settlements were narrow crooked lanes, imitation medieval walls and gates. Furthermore, car ownership was so low that most architects simply may not have seen any reason to build streets which were wide enough for car access or parking and the economic circumstances made it more necessary to build as cheaply as possible, which meant narrow streets and very few major roads.

One of the most important planners during the Weimar Republic was Hermann Jansen, who developed a street design which would integrate the tradition of the garden suburbs with the modern needs of motor traffic. His independent footpath network made it possible to reach the shops, the school and kindergarten without crossing any major streets. Of crucial importance again were the communal greens which were interwoven into the housing estate, often axial in form, to serve as pedestrian 'streets'. Many German architects copied Jansen's design. By the beginning of the 1920s, his street designs look amazingly similar to Radburn, though over- and underpasses were not built because of the far lower number of cars. Even so, although Radburn principles were highly praised in Germany, it seems likely that they were the product of independent American thought, but were evidently influenced by British (and German) planners.

During the 1930s and '40s, the plans of the newly built residential estates in Germany increasingly became replicas of the garden suburbs of the pre-World War I period. This was also expressed in the street networks, which favoured narrow streets and independent footpaths. The planned new towns included wide green axes which would come far into the city centre and connect the densely built housing areas with the outskirts. Again we can only speculate why such planning was possible during the Nazi period. One explanation, which is hinted at in the literature, could be that Hitler simply was not interested in settlement planning, and thus street designs did not take account of Hitler's desire to lead Germany into the new age of mass motorization.

Several of the redevelopment plans of post-war Germany continued to use green axes but it was then condemned as 'Nazi design' and was only used on a smaller scale in some housing estates.

When, during the 1970s, Germany went back to many of the 'old' ideas of traffic calming, reinterpreting them as a 'new' concept of protecting pedestrians from motor traffic, motor traffic had already severely affected and damaged urban areas and, even more important, the damage was visible to many people. The Dutch *Woonerven* and Buchanan's concept of environmental areas were then widely seen as presenting an alternative approach. However it would be erroneous to say that traffic calming was copied from the Netherlands or from Britain; it has to be seen primarily as a reaction against too much road-building, too many cars and as a further development of the success of pedestrianization.

Why, it may be asked, is Britain today still lukewarm in adopting traffic calming as a policy if it is indeed part of its own historical evolution? The answer may lie in the fact that in Britain, for various reasons, nothing like the scale of German road-building or the degree of motorization have been achieved. British roads – including residential roads – are flooded with motor vehicles and its traffic engineers have become masters at making the most of any available road space capacities, but there are limits to their skills. In the mean time, the urban environment has in many cases deteriorated badly although it was for a long time not seen as an important issue for which new alternative policies were needed. The late 1980s has experienced a growing British 'green' movement which would very much like to implement traffic calming, but such policies in Britain may often imply a reduction in the number of motor-vehicle trips because of the lack of road space. A limited increase in road space in or near British towns is almost certainly needed, especially in the form of bypasses, but great care must be taken that it is not overdone. In fact, Britain is far luckier than Germany because of its lack of previous road-building but as a result it needs much more radical traffic calming measures than its continental neighbour(s). It must also seek to remedy the long-standing neglect of public and freight transport which will become even more crucial in future.

The overall theme of this book has been to consider how Britain and Germany reacted to and coped with the motor vehicle. The conflict between the pedestrian and the motor vehicle was anticipated by many very early on. Policies were suggested and partly implemented but none of the very expensive and rather fundamental changes was ever made. Yet, however impressive the history of protecting the pedestrian may appear, in reality it has only been thus far a very limited attempt. Even policies such as large-scale pedestrianization or area-wide traffic calming carried out in Germany since the beginning of the 1980s have been limited in scale when compared with the whole city. Only very recently do we see in Germany the attempt to traffic-calm whole cities, but despite that there are still many features which show a half-hearted approach, such as the low taxation of car use or the lack of public-transport promotion.

If we evaluate these policies in the context of both the continuous increase in car use and the change in the urban environment then we can only conclude that the protection of the pedestrian has been addressed in a much more limited way than has any of the other socio-economic or technical problems modern societies have been faced with. There is hardly another invention, which has, as much as the motor vehicle, become at the same time both a nuisance and a symbol of progress, and has been allowed to continue to destroy life and health in such large numbers, together with valuable parts of our urban heritage. Without doubt, motorization has a unique position in society which has been pointed out by many writers. Perhaps, the most prominent of them, Sir Colin Buchanan, wrote 'in the hundred years of its existence the motor vehicle has been a constant source of anxiety, love, hatred, condemnation and political bickering, and we don't seem to be anywhere near the end of it'.

In the final analysis, each society will have to choose its own strategy. If it wants to preserve the environment, then it has to calm the car far more than it has done in the past. If that is seen, however belatedly, as a high priority then the pedestrians and the residents will begin to be properly protected.

References

Abel, A. (1942) Grundsätzliches im Städtebau an Hand einer Reihe von Planungen, II. Zurückgewinnung der Verkehrsflächen für den Fußgänger, *Monatshefte für Baukunst und Städtebau*, Vol. 26, 10, pp. 221–32.

Abel, A. (1950) *Regeneration der Städte*, Zürich, Verlag für Architektur.

Abercrombie, P. (1924) *Sheffield: A Civic Survey and Suggestions towards a Development Plan*, London, University Press of Liverpool.

Abercrombie, P. (1945a) *A Plan for Bath*, Bath, Pitman.

Abercrombie, P. (1945b) *Greater London Plan 1944*, London, HMSO.

Abercrombie, P. and B. F. Brueton (1930) *Bristol and Bath Regional Planning Scheme*, London, The University Press of Liverpool.

Abercrombie, P. and A. C. Holliday (1921) New Arterial Roads in Course of Construction, *The Town Planning Review*, Vol. 9, pp. 67–72.

Abercrombie, P. and J. H. Forshaw (1943) *County of London Plan*, London, Macmillan.

Adams, J. (1985) *The Risk and Freedom*, London, Bottesford Press.

Adshead, S. D. (1911) Review on Robinson's book, *The Town Planning Review*, Vol. 2, p. 330.

Adshead, S. D. (1913) The Third International Road Congress, *The Town Planning Review*, Vol. 4, pp. 234–44.

Adshead, S. D. (1915) The Layout of Roads in Relation to Requirements, *The Town Planning Review*, Vol. 6, pp. 163–70.

Adshead, S. D. (1923) *Town Planning and Town Development*, London, Methuen.

Adshead, S. D. (1941) *A New England*, London, Muller.

Adshead, S. D. (1943) *New Towns for Old*, London, Letchworth, Dent.

Adshead, S. D., C. J. Minter and C. W. C. Needham (1948) *York – A Plan for Progress and Preservation*, York.

Apel, D. (1971) *Ein Beitrag zur Bewertung der von Kraftfahrzeugverkehr beeinflußten Umweltqualität von Stadtstraßen*, Dissertation, Rheinisch-Westfälische Technische Hochschule Aachen, Aachen.

Appleyard, D. (1981) *Livable Streets*, Berkeley, University Press.

Ashworth, W. (1954; 1965) *The Genesis of Modern British Town Planning*, London, Routledge & Kegan Paul.

Baier, R., H. Collin, H. H. Keller, P. Müller *et al.* (1989) *Wirkung von Tempo 30-Zonen*, Darmstadt, Ad-hoc-Gruppe 'Geschwindigkeitsbekämpfung'.

Balchin, J. (1980) *First New Town: An Autobiography of the Stevenage Development Corporation 1946–1980*, Stevenage, Stevenage Development Corporation.

Bardou, J-P, J-J. Chanaron *et al.* (1982) *The Automobile Revolution. The Impact of the Industrie*, Chapel Hill, University of North Carolina.

Bauer, C. (1935) *Modern Housing*, London, Allen & Unwin.

Baumeister, R. (1876) *Stadterweiterungen*, Berlin, Ernst & Korn.

Baumeister, R. (1887) *Moderne Stadterweiterungen*, Hamburg, Richter.

Baynes, N. H. (1942) *The Speeches of Adolf Hitler*, 2 Vols, London, New York, Toronto, Oxford University Press.

Beesley, M. E. and J. F. Kain (1964) Urban Form Car Ownership and Public Policy: An Appraisal of Traffic in Towns, *Urban Studies*, Vol. 1, pp. 174–203.

Beesley, M. E. and J. G. Roth (1962) Restraint of Traffic in Congested Areas, *The Town Planning Review*, Vol. 33, pp. 184–96.

Behn, O. (1983) *Passagen in der Hamburger City – Eine Empirische Untersuchung Ihrer Benutzer*, Institute für Soziologie der Universität Hamburg und Gerellschaft für sozialwissenschaftliche Stadtforschung, Hamburg.

Behrendt, C. W. (1913) Großstädtische Wohnquartiere, *Bau-Rundschau*, Vol. 4, pp. 101–9.

Berlepsch-Valendás, B. D. A. (1912) *Die Gartenstadtbewegung in England, ihre Entwickelung und ihr jetziger Stand*, München, Berlin, Oldenburg.

Berlepsch-Valendás, B. D. A. (1913) Projekt Gartenstadt Obereßlingen, *Gartenstadt*, Vol. 7, pp. 84–91.

Bernoulli, H. (1911) Die neue Stadt, *Gartenstadt*, Vol. 5, pp. 109–12.

Bernoulli, H. (1954) Die Fußgängerstadt, *Baukunst und Werkform*, Heft 6, und Plan 11, pp. 27–9.

BfLR, BASt, UBA (1983) Flächenhafte Verkehrsberuhigung Zwischenbericht, *Informationen zur Raumentwicklung*, Heft 8/9.

BfLR, BASt, UBA (1985) *3. Kolloquium, Forschungsvorhaben 'Flächenhafte Verkehrsberuhigung'* Erste Erfahrungen aus der Praxis, 30. September – 1. Oktober, Berlin.

BfLR, BASt, UBA (1988) *4. Kolloquium, Forschungsvorhaben 'Flächenhafte Verkehrsberuhigung'* Ergebnisse aus drei Modellstädten, 26.–27. Mai, Buxtehude.

BfLR, BASt, UBA (1988) *Verkehrsberuhigung und Entwicklung von Handel und Gewerbe*, Materialien zur Diskussion, Bonn.

BMBau (1973) *Raumordnungsbericht 1972*, Bonn.

BMBau (1974) *Die Rolle des Verkehrs in Stadtplanung, Stadtentwicklung und Städtische Umwelt*, Bonn.

BMBau (1978a) *Ausländische Erfahrungen mit Möglichkeiten der räumlichen und sektoralen Umverteilung des städtischen Verkehrs*, Bonn.

BMBau (1978b) *Siedlungsstrukturelle Folgen der Einrichtung verkehrsberuhigter Zonen in Kernbereichen*, Bonn.

BMBau (1979) *Verkehrsberuhigung. Ein Beitrag zur Stadterneuerung*, Bonn.

BMBau (1980) *Wohnstraßen der Zukunft*, Bonn.

BMBau (1982) *Planungsfibel zur Verkehrsberuhigung*, Bonn.

BMBau (1983) *Kostenhinweise zur Verkehrsberuhigung*, Bonn.

BMBau (1985) *Verkehrsberuhigung und Stadtverkehr*, Bonn.

BMBau (1986) *Stadtverkehr im Wandel*, Bonn.

Boeminghaus, D. (1982) *Fußgängerbereiche und Gestaltungselemente*, Stuttgart, Krämer.

Bongard, A. E. and I. C. Bongard (1983) Die Ausbildung und Prüfung von Fahrlehrern in der Bundesrepublik Deutschland, *Unfallforschung- und Sicherheitsforschung, Straßenverkehr*, Vol. 43, pp. 1–21.

Bor. W. G. (1963) Environmental Approach to Urban Traffic, *RIBA Journal*, Vol. 70, pp. 135–64.

Bourgin, C., P. H. Bovy, and D. Teasdale (1979) Besancon, in OECD (ed.), *Managing Transport*, Paris, pp. 43–66.

Bowers, P. H. (1986) Environmental Traffic Restraint: German Approaches to Traffic Management by Design, *Built Environment*, Vol. 12, pp. 60–73.

Bowers, P. H. and J. C. Thomson (1984) Shared Space and Child Safety, *Traffic Engineering and Control*, Vol. 25, pp. 550–4.

Bramilla, R. and G. Longo (1974) *The Rediscovery of the Pedestrian, 12 European Cities*, Institute for Environmental Action and Columbia University Centre for Advanced Research in Urban and Environmental Affairs, Washington D.C.

Bramilla, R. and G. Longo (1977) *For Pedestrians Only*, New York, Watson-Guptill Publication.

Brandenburg, Dr Ing. (1936) Kraftverkehrswirtschaft, *Zeitschrift für Verkehrswissenschaft*, Vol. 13, pp. 89–104.

Breines, S. and W. J. Dean (1974) *The Pedestrian Revolution – Streets without Cars*, New York, Vintage.

British Road Federation (1989) *Basic Road Statistics 1989*, London.

Brix, E. (1954) Städtebau der Zukunft, *Verkehr und Technik*, Vol. 7, pp. 158–9.

Brodie, J. A. (1914) Some Notes on the Development of Wide Roads for Cities, *The Town Planning Review*, Vol. 5, pp. 294–9.

Brownell, B. A. (1980) Urban Planning, the Planning Profession, and the Motor Vehicle in Early Twentieth-century America, in G. E. Cherry (ed.), *Shaping an Urban World*, London, Mansell.

Buchanan, C. D. (1956) The Road Traffic Problem in Britain, *The Town Planning Review*, Vol. 26, pp. 215–41.

Buchanan, C. D. (1958) *Mixed Blessing, The Motor in Britain*, London, Leonard Hill.

Buchanan, C. D. (1960) Transport – The Crux of City Planning, *RIBA Journal*, Vol. 68, pp. 69–74.

Buchanan, C. D. (1961) Standards and Values in Motor Age Towns, *Journal of the Town Planning Institute*, Vol. 47, pp. 320–9.

Buchanan, C. D. (1963) Traffic in Towns, in British Road Federation (ed.), *People and Cities, Report of the 1963 London Conference*, pp. 15–22.

Buchanan, C. D. (1964) The Urban Environment, *Journal of the Town Planning Institute*, Vol. 50, pp. 268–74.

Buchanan, C. D. (1988) A Historical Review of the 'Traffic in Towns' Report of 1963, *PTRC Transport and Planning Summer Annual Meeting*, University of Bath, 12–16 September 1988 (unpublished).

Buchanan, C. and Partners (1965) *Bath: A Planning and Transport Study*, London.

Buchanan, C. and Partners (1966) *Cardiff: Development and Transportation Study*, Report of the Probe Study, London.

Buchanan, C. and Partners (1968) *Cardiff: Development and Transportation Study*, Main Study Report, London.

Buchanan, C. and Partners (1980) *Transport Planning for Greater London*, Westmead, Farnborough, Saxon House.

Buchanan, M. (1988) Urban Transport and Market Forces in Britain, in C. Hass-Klau (ed.), *New Life for City Centres*, London, Anglo-German Foundation, pp. 211–19.

Burnham, D. H. (1906) *Report on a Plan for San Francisco* (reprinted 1971), Berkeley, Urban Books.
Burnham, D. H. (1910) A City of the Future under a Democratic Government, in The Royal Institute of British Architects (ed.), *Transactions of the Town Planning Conference*, London, pp. 368–78.
Büekschmitt, J. (1963) *Ernst May, Bauen und Planung 1*, Stuttgart, A. Koch.
Cashmore, J. F. (1981) Towards an Evaluation Method for Pedestrianisation Schemes, final year dissertation, Diploma in Town and Regional Planning, Leeds Polytechnic.
Chambless, E. (1910) *Roadtown*, London, Sidgwick & Jackson.
Cherry, G. E. (1974) *The Evolution of British Town Planning*, Leighton Buzzard, Leonard Hill.
Choay, F. (1969) *The Modern City: Planning in the 19th Century*, London, Studio Vista.
Churchill, H. (1983) Henry Wright: 1878–1936, in D. A. Krueckeberg (ed.), *The American Planner Biographies and Recollections*, New York, London, Methuen.
Condit, C. W. (1973) *Chicago 1910–1929, Building, Planning and Urban Technology*, Chicago, London, University of Chicago Press.
Corbusier, Le M. (1924) *Urbanisme*, Edition Cres, Paris; UK edn (1971), *The City of To-morrow*, London, Architectural Press.
Corbusier, Le M. (1933) *The Radiant City*, London, Faber & Faber, republished 1964.
Creese, W. L. (1966) *The Search for Environment, The Garden City: Before and After*, New Haven, London, Yale University Press.
Creese, W. L. (1967) *The Legacy of Raymond Unwin: A Human Pattern for Planning*, Cambridge, Mass., London, MIT Press.
Crowther, G. (1963) Discussion: 'Traffic in Towns', in British Road Federation (ed.), *People and Cities, Report of the 1963 Conference*, pp. 23–6.
DAfSL (Deutsche Akademie für Stadt- und Landesplanung) *Deutscher Städtebau nach 1945*, Hannover.
Dalby, E. (1988) Self-Enforcing Systems for Controlling Traffic Speed in Urban Areas – A Survey of British Experience, Department of Transport (not published).
Dahl, R. (1972) *Der Anfang vom Ende des Autos*, Münden, Langewiesche Brandt.
Department of the Environment, Department of Transport (1977) *Design Bulletin 32, Residential Roads and Footpaths Layout Considerations*, London, HMSO.
Department of Transport (1987a) *Road Accidents Great Britain 1986*, the Casualty Report, London, HMSO.
Department of Transport, Transport Advisory Unit Leaflet (1987b) *Measures to Control Traffic for the Benefit of Residents, Pedestrians and Cyclists*, London, HMSO.
Department of Transport (1988) *Getting the Right Balance*, Local Transport Note 1/87, Welsh Office, London, HMSO.
Der Bundesminister für Raumordnung, Bauwesen und Städtebau (1988) *Städtebauliche Integration von innerörtlichen Hauptverkehrsstraßen, Städteumfrage*, Bonn.
Der Bundesminister für Verkehr (1975) *Verkehr in Zahlen 1975*, Bonn.

Der Bundesminister für Verkehr (1984) *Geschichte der deutschen Straßenverkehrszeichen*, Bonn.

Der Bundesminister für Verkehr (1988) *Verkehr in Zahlen 1988*, Bonn

Der Bundesminister für Verkehr, Allgemeiner Deutscher Automobil Club (ADAC) (1977) *Sicherheit für den Fußgänger II Verkehrsberuhigung, Erfahrungen und Vorschläge in Städten und Gemeinden und Schlußfolgerungen aus dem Städtewettbewerb 1977*, Bonn, München.

Der Minister für Wirtschaft, Mittelstand und Verkehr des Landes Nordrhein-Westfalen (1979) *Großversuch 'Verkehrsberuhigung in Wohngebieten'*, Köln, Kirschbaum.

Deselaers, Regierungsrat (1955) Neuverteilung der Rechte von Kraftverkehr, Fußgänger und Straßenanlieger, *Der Städtetag*, pp. 509–11.

Dollinger, H. (1972) *Die totale Autogesellschaft*, Passau, Carl Hanser.

Dupree, H. (1987) *Urban Transport: The New Town Solution*, Aldershot, Gower.

Ehlgötz, H. (1925) Der Einfluß des Kraftfahrzeugverkehrs auf Städtebau und Siedlungswesen, *DBZ, Stadt und Siedlung*, Vol. 59, pp. 44–8, and pp. 70–2.

Ehlgötz, H. (1928) Großstadt und Verkehrsregelung, aus der Ausstellung: 'Die technische Stadt' in Dresden, *DBZ, Stadt und Siedlung*, Vol. 62, pp. 117–26.

Eichenauer, M., von Winning, H.-H., Streichert, E. (1978) *Sicherheit und Verhalten in Verkehrsberuhigten Zonen*, BASt, Köln, published again in 1980.

Eichenauer, M., Streichert, E., von Winning, H.-H. (1980) *Innerstädtische Planung als Einflußgröße der Verkehrssicherheit*, BASt, Köln.

Engel, H. (1976) Stadtgebiete des 19. Jahrhunderts und ihre denkmalpflegerischen Probleme am Beispiel Berlins. Entwicklung der städtebaulichen Grundvorstellungen seit dem 17. Jahrhundert, in H. Engel (ed.), *Stadtidee und Stadtgestaltung*, Berlin, Abakon-Verlag.

Eno, W. P. (1929) *Simplification of Highway Traffic*, Sangatuck, Fairfield, Connecticut, Eno Foundation for Highway Traffic Regulation.

Fehl, G. (1983) The Niddatal Project – The Unfinished Satellite Town on the Outskirts of Frankfurt, *Built Environment*, Vol. 9, pp. 185–97.

Fehl, G. Rodriguez-Lores, J. (1983) Die 'Gartenstadt-Bebauung', *Bauwelt*, No. 12, pp. 462–71.

Fest, J. C. (1974) *Hitler*, London, Weidenfeld & Nicolson.

Fiebig, K.-H., B. Horn *et al.* (1988) Umweltverbesserung in den Städten, *Difu*, Heft 5: Stadtverkehr, Berlin.

Filler, R. (1986) *A History of Welwyn Garden City*, Chichester, Phillimore.

Först, W. (1962) *Chaos oder Ordnung auf unseren Straßen*, Köln, Deutscher Städtetag.

Ford, G. B. (1920) *L' Urbanisme en Practique*, Paris, Ernest Leroux.

Foster, C. D. (1962) Electronics and Automation on the Road, *Road International, Conference*, Summer 1962.

Frank (1934) Not officially published – letter written to local authority in Hamburg.

Fritsch, T. (1912) *Die Stadt der Zukunft*, Leipzig, Hammer.

Garbrecht, D. (1978) Fußgängerbereiche – ein Alptraum? *BAG-Nachrichten*, No. 8, pp. 3–5, first published 1977.

Garbrecht, D. (1979) Zu Fuß gehen – Rückschritt oder Fortschritt?, in F. Duve (ed.), *Technologie und Politik, Das Magazin zur Wachstumskrise*, Vol. 14, pp. 226–41.

Garbrecht, D. (1981) *Gehen, Plädoyer für das Leben in der Stadt*, Weinheim, Basel, Beltz Verlag.

Gantvoort, J. Th. (1982) Practice in North and West Europe: Netherlands, University of Strathclyde: Department of Urban & Regional Planning in Association with Continuing Education Office, Seminar: *World Development in Pedestrian Planning*: 1–2 July (unpublished).

Geist, J. F. (1983) *Arcades: The History of a Building Type*, Cambridge, Mass., London, MIT Press.

Geuns, van, L. A. (1981) Pedestrian Barriers and Benefits: Experiences in the Netherlands, *Journal of Urban and Environmental Affairs*, Vol. 13, pp. 121–34.

GLC (1973) GLC Study Tour of Europe and America, *Pedestrianised Streets*, London.

Goecke, T. (1915) Der Kleinwohnungsbau die Grundlage des Städtebaues, *Der Städtebau*, Vol. 12, pp. 3–5.

Goecke, T. (1918) Der Bebauungsplan der Großstadt, in C. J. Fuchs (ed.), *Die Wohnungs- und Siedlungsfrage nach dem Kriege*, Wilhelm Meyer-Ilschen, Stuttgart, pp. 155–68.

Goudappel, H. M. and P. V. von Gurp (1974) Dutch Town Centres, in OECD (ed.), *Streets for People*, Paris, pp. 113–21.

Grahame, K. (1908) *Tales from the Wind in the Willows*, Harmondsworth, Penguin.

Gregory, T. (1973) Coventry, in J. C. Holliday (ed.), *City Centre Redevelopment: A Study of British City Centre Planning and Case Studies of Five English City Centres*, London, Charles Knight, pp. 78–134.

Gunkel, F. (1965) Verkehrsprobleme in Verdichtungsgebieten aus der Sicht des Landesplaners, *Verkehr und Raumordnung*, Vol. 35, pp. 7–26.

Gurlitt, C. (1929) New Yorker Siedlungen, *Stadtbaukunst*, Vol. 10, pp. 27–31.

Hajdu, J. (1983) Postwar Development and Planning of West German Cities, in T. Wild (ed.), *Urban and Rural Change in West Germany*, London & Canberra, Beckenham, Croom Helm, pp. 16–39.

Hall, P. (1980) *The Great Planning Disaster*, London, Weidenfeld & Nicolson.

Hall, P. (1984) Metropolis 1890–1940: Challenges and Responses, in A. Sutcliffe (ed.), *Metropolis 1890–1940*, London, Mansell, pp. 19–66.

Hall, P. and C. Hass-Klau (1985) *Can Rail Save the City?*, Aldershot, Gower.

Hall, P. (1988) *Cities of Tomorrow, An Intellectual History of Urban Planning and Design in the Twentieth Century*, Oxford, New York, Basil Blackwell.

Hamer, M. (1987) *Wheels within Wheels – A Study of the Road Lobby*, London, Routledge & Kegan Paul.

Hansard (1933–4) *Local Acts*, Ch. XCV ii, London.

Hartmann, K. (1976) *Gartenstadtbewegung*, München, Heinz Moos.

Hartwig, A. (1935) Hans Güldenpfennig: Kölner Verkehrsprobleme und Domumbau, *Die Baugilde*, Vol. 17, pp. 701–9.

Hass-Klau, C. H. M. (1982) 'New' Transport Technologies in the Federal Republic of Germany, *Built Environment*, Vol. 8, pp. 190–6.

Hass-Klau, C. (1984) German Urban Public Transport Policy, *Cities*, Vol. 1, pp. 551–6.

Hass-Klau, C. (1986) Environmental Traffic Management: Pedestrianisation and Traffic Restraint – A Contribution to Road Safety, *PTRC Transport Policy*, pp. 91–104. London.

Hass-Klau, C. (1989) International Experience of Traffic Calming: The Solution to British Transport Problems?, in Traffex'89 Conference (ed.), *Road Safety and Traffic Calming*, 6 April 1989, PTRC, London.

Heckscher, A. and P. Robinson (1977) *Open Spaces: The Life of American Cities*, New York, San Francisco, Harper & Row.

Hegemann, W. (1925) *Amerikanische Architektur und Stadtbaukunst*, Berlin, Ernst Wasmuth.

Hegemann, W. (1930) *Das steinerne Berlin: Geschichte der größten Mietskasernenstadt der Welt*, Berlin, Kiepenheuer.

Hegemann, W. (1933) Südamerikanische Verkehrsnöte, *Monatshefte für Baukunst und Städtebau*, Vol. 27, pp. 89–94.

Heilig, W. (1935) Der Kraftwagen, ein Bahnbrecher für den Umbau der Großstadt, *Die Straße*, Vol. 2, pp. 738–41.

Heller, K. (1928) Das Kraftwagenstraßennetz Deutschlands, *Verkehrstechnik*, Vol. 9, pp. 669–72.

Hénard, P. E. (1910) Les Villes de l'Avenir, *Royal Institute of British Architects, Transaction of the Town Planning Conference*, Oct. 1910, pp. 345–67.

Hilberseimer, L. (1929) Die Neue Geschäftsstraße, *Das neue Frankfurt*, Vol. 3, pp. 67–77.

Hilberseimer, L. (1944) *The New City, Principles of Planning*, Chicago, Paul Theobald.

Hilberseimer, L. (1955) *The Nature of Cities, Origin, Growth and Decline, Pattern and Form, Planning Problems*, Chicago, Paul Theobald & Co.

Hillebrecht, R. (1953) Ein Deutscher Städtebauer sieht Amerika, *Verkehr und Technik*, Vol. 6, p. 191.

Hillebrecht, R. (1957) Verpaßte Chance der Städtebauer, *Der Städtetag*, Vol. 10, pp. 69–72.

Hillebrecht, R. (1965) Stadt Hannover: Die Planung des Wiederaufbaus der Stadt nach 1945, *Urbanistica*, Vol. 53, pp. 22–40.

Hillman, M. and A. Whalley (1979) Walking *is* Transport, *Policy Study Institute* (PSI), Vol. 45, No. 583, London.

Himmler, H. (1942) Richtlinien für die Planung und Gestaltung der Städt e in den eingegliederten deutschen Ostgebieten, in A. Teut (1967, ed.), *Architektur im Dritten Reich 1933–1945*, Berlin, Frankfurt/Main, Wien, Ullstein, pp. 347–57.

Hines, T. S. (1974) Burham of Chicago Architect and Planner, New York, Oxford University Press.

Hoffmann, R. (1961) *Die Gestaltung des Verkehrswegenetzes*, Hannover, Jänecke.

Hofmann, W. (1987) in Hermann Jansen, W. Ribbe, W. Schäche (eds), *Baumeister Architekten Stadtplaner*, Berlin, Historische Kommission zu Berlin.

Hollatz, J. W. (1954) Der Verkehr als Faktor der städtebaulichen Gestaltung, *Verkehrswissenschaftliche Veröffentlichungen*, Heft 3, Stadtverkehr heute und morgen, Düsseldorf, Droste Verlag, pp. 45–68.

Hollatz, J. W. and F. Tamms (1965, eds) *Die kommunalen Verkehrsprobleme in der Bundesrepublik Deutschland*, Essen, Vulkan.

Horadam, M. C. (1983) Verkehrsordnungen und Verkehrssicherheit – Rückblick auf deutsche Straßenverkehrsordnungen von Beginn bis heute, Bundesanstalt für Straßenwesen (BASt), Köln (unpublished).

Horsfall, T. C. (1904) *The Improvement of the Dwellings and Surroundings of the People, the Example of Germany*, Manchester, University Press.

Jackson, F. (1985) *Sir Raymond Unwin, Architect, Planner and Visionary*, London, A Zwemmer Ltd.

Jansen, H. (1917) *Die Großstadt der Neuzeit*, Konstantinopel, Ahmed Ihsan & Co.

Kahmann, H. (1979) Modell für Deutschland? *Arch +*, Vol. 47, pp. 33–7.

Kampffmeyer, B. (1910) Unsere soziale Studienreise nach England, *Gartenstadt*, Vol. 5, pp. 80–4.

Kampffmeyer, H. (1909) *Die Gartenstadtbewegung*, Leipzig.

Kampffmeyer, H. (1918) Die Gartenstadtbewegung, in C. J. Fuchs (ed.), *Die Wohnungs- und Siedlungsfrage nach dem Kriege*, Meyer-Ilschen, Stuttgart, pp. 331–49.

Kantor, H. (1983) Charles Dyer Norton and the Origins of the Regional Plan of New York, in D. A. Krueckeberg (ed.), *The American Planner, Biographies and Recollection*, New York, Methuen, pp. 179–95.

Kautt, D. (1983) *Wolfsburg im Wandel städtebaulicher Leitbilder*, Wolfsburg, Stadt Wolfsburg.

Keller, H. H. (1981) Verkehrsberuhigung in Unterhaching und in Berlin-Charlottenburg, *Bauverwaltung*, Heft 9, pp. 351–6.

Keller, H. H. (1987) Verkehrsberuhigung könnte mit Tempo 30 beginnen, *Der Deutsche Städtetag*, Heft 1, pp. 2–7.

Kellett, J. R. (1969) *The Impact of Railways on Victorian Cities*, London, Routledge & Kegan Paul.

Kiepe, F. (1988) Innerörtliche Geschwindigkeitsbegrenzungen und Verkehrsberuhigung, Vorbericht für die 34. Sitzung der Konferenz leitender kommunaler Verkehrsplaner am 9/10.6. 1988 in Hamburg, *Deutscher Städtetag*, Köln.

Klapheck, R. (1930) *Siedlungswerk Krupp*, Berlin, Wasmuth.

Klessmann, E. (1981) *Geschichte der Stadt Hamburg*, Hamburg, Hoffmann & Campe.

Koller, P. (1939) Die Stadt des KDF-Wagens, Sonderdruck aus der Zeitschrift 'Die Kunst im Deutschen Reich', München, Franz Eher.

Koller, P. (1940) Die Siedlung Steimkerberg im Rahmen der Stadtplanung, *Bauen, Siedeln, Wohnen*, Vol. 20, pp. 656–61.

Korte, J. M. (1958) *Grundlagen der Straßenverkehrsplanung in Stadt und Land*, Wiesbaden, Berlin, Bauverlag GmbH.

Koller, P. (1987) Unpublished correspondence.

Landeshauptstadt München, Referat für Stadtplanung und Bauordnung (1987) *Arbeitsberichte zur Stadtentwicklungsplanung, Planungsgrundlagen für die Münchner Innenstadt*, München.

Leibbrand, K. (1964) *Verkehr und Städtebau*, Basel, Stuttgart, Birkhäuser Verlag.

London County Council (1951) *Development Plan 1951*, Analysis, London.

LPAC (London Planning Advisory Committee) (1988) *Strategic Planning Advice for London Policies for the 1990's*, Draft, London.

London Traffic Branch of the Board of Trade (1910, 1911) *Houses of Parliament, Reports from Commissioners, Inspectors and Others*, Vol. 34, London.
Lubove, R. (1963) *Community Planning in the 1920s: The Contribution of the Regional Planning Association of America*, Pittsburgh: Pittsburgh University Press.
Ludmann, H. (1972) *Fußgängerbereiche in Deutschen Städten*, Köln, Deutscher Gemeindeverlag, W. Kohlhammer.
MacKay, D. H. and A. W. Cox (1979) *The Politics of Urban Change*, London, Croom Helm.
MacKaye, B. (1930) The Townless Highway, *The New Republic*, Vol. 62, pp. 93–5.
Mäcke, A. (1977) Stadt Region Land, *Schriftenreihe des Instituts für Stadtbauwesen*, Rheinisch-Westfälische Technische Hochschule Aachen, Aachen.
Mäcke, P. (1954) Grundlagen und Grundelemente der Verkehrsanlagen für den motorisierten Stadtverkehr, Ministerium für Wirtschaft und Verkehr des Landes Nordrhein-Westfalen (ed.), *Stadtverkehr heute und morgen, Verkehrswissenschaftliche Veröffentlichungen*, Heft 31, Düsseldorf, Droste Verlag, pp. 117–39.
Marples, E. (1963) Introduction, in British Road Federation (ed.), *People and Cities, Report of the 1963 London Conference*, pp. 11–14.
Martin, H. E. L. (1915) Roads, *The Town Planning Review*, Vol. 6, pp. 57–63.
May, E. (1930) Fünf Jahre Wohnungsbautätigkeit in Frankfurt am Main, *Das neue Frankfurt*, Vol. 4, pp. 21–70.
May, E. (1963) Die Verstopfte Stadt, *Bauwelt*, Heft 54, pp. 183–4.
Matzerath, H. (1984) Berlin, 1890–1940, in A. Sutcliffe (ed.), *Metropolis 1890–1949*, London, Mansell, pp. 289–318.
Maxcy, G. and A. Silberston (1959) *The Motor Industry*, London, Allen & Unwin.
Meadows, D. L. (1972) *The Limits to Growth*, New York, Universe Books.
Menke, R. (1977) Verkehrsplanung für wen?, *Stadtbauwelt*, Heft 53, pp. 19–23 (361–5).
Miller, M (1982) Der rationale Enthusiast, Raymond Unwin als Bewunderer deutschen Städtebaues, *Stadtbauwelt*, Heft 75, pp. 319–26 (1513–20).
Ministry of Transport (1920) *Report of the Advisory Committee on London Traffic*, London, HMSO.
Ministry of War Transport (1946) *Design and Layout of Roads in Built-Up Areas*, London, HMSO.
Ministry of Transport (1963) *Traffic in Towns: A Study of the Long Term Problems of Traffic in Urban Areas*, London, HMSO.
Ministry of Transport (1964) *Road Pricing: The Economic and Technical Possibilities* (Chairman: R. Smeed), London, HMSO.
Ministry of Transport (1967a) *Cars for Cities, A Study of Trends in the Design of Vehicles with particular Reference to their Use in Towns*, London, HMSO.
Ministry of Transport (1967b) *Better Use of Town Roads*, London, HMSO.
Minister für Wirtschaft, Mittelstand und Verkehr des Landes Nordrhein-Westfalen (1977) *Verkehrsberuhigung in Wohngebieten*, Köln.
Minuth, K-H. (1983) Die Regierung Hitlers, Teil 1: 1933/34, Vol 1 + 2, K. Repgen (ed.), *Akten der Reichskanzlei Regierung Hitler 1933–1938*, Boppard am Rhein, Harald Boldt.

Mitscherlich, A. (1969) *Die Unwirtlichkeit userer Städte Antsiftung zum Unfrieden*, Frankfurt am Main, Suhrkamp.

Monheim, H. (1979) Verkehrsberuhigung durch Stadtschnellstraßen, *Arch +*, Vol 47, pp. 10–13.

Monheim, R. (1975) *Fußgängerbereiche*, Köln, Deutscher Städtetag.

Monheim, R. (1980) *Fußgängerbereiche und Fußgängerverkehr in Stadtzentren in der Bundesrepublik Deutschland*, Bonn, Dümmlers.

Monheim, R. (1986) Pedestrianization in German Towns: A Process of Continual Development, *Built Environment*, Vol. 12, pp. 30–43.

Monheim, R. (1987) Entwicklungstendenzen von Fußgängerbereichen und verkehrsberuhigten Einkaufsstraßen, *Arbeitsmaterialien zur Raumordnung und Raumplanung*, Heft 41.

Monheim, R. (1988) Pedestrian Zones in West Germany – The Dynamic Development of an effective Instrument to enliven the City Centre, in C. Hass-Klau (ed.), *New Life for City Centres*, Anglo-German Foundation for the Study of the Industrial Society, Bonn, London, pp. 107–56.

Morlock, G. (1987) Öffentlichkeitsarbeit zur Zonen-Geschwindigkeits-Beschränkung, *Der Städtetag*, pp. 593–6.

Müller, P. (1971) Fußgängerverkehr in Wohnsiedlungen, Bundesminister für Verkehr (ed.), *Straßenbau und Straßenverkehrstechnik*, Bonn.

Müller, P. (1985) *Geschwindigkeitsverhalten*, Darmstadt, Institut Wohnen und Umwelt.

Müller, P. (1986) Tempo 30 – Modellversuche, Darmstadt, Institut Wohnen und Umwelt.

Müller, P. and H. H. Topp (1986) Verkehrsberuhigung durch Straßenumbau: Eine neue Art der Stadtzerstörung? *Deutscher Städtetag*, pp. 327–30.

Mulzer, E. (1972) *Der Wiederaufbau der Altstadt von Nürnberg*, Fränkische Geographische Gesellschaft, Erlangen, Heft 31.

Mumford, L. (1961) *The City in History*, Harmondsworth, Penguin.

Muthesius, H. (1979) *The English House*, London, Crosby Lockwood Stables (Das englische Haus was originally published in 1904, 1905 by Wasmuth, Berlin).

Napp-Zinn, A. F. (1933) Zur Verkehrspolitik des national-sozialistischen Staates, *Zeitschrift für Verkehrswissenschaften*, Vol. 11, pp. 77–90.

National Consumer Council (1987) *What's Wrong with Walking?* London, HMSO.

Nelson, W. H. (1970) *Small Wonder The Amazing Story of the Volkswagen*, London, Hutchinson, first edition 1967.

Niehusener, W. (1974) Essen, Germany, in OECD (ed.), *Streets for People*, Paris, pp. 71–8.

North, J. (1983) Developments of Transport, in T. Wild (ed.), *Urban and Rural Change in West Germany*, London & Canberra, Beckenham, Croom Helm. pp. 130–60.

OECD (1979) *Traffic Safety in Residential Areas*, Paris.

Olmsted, F. L. (Jr) and T. Kimball (1922; 1928) *Olmsted, F. L. (Sr.) Landscape Architect 1822–1903*, Vol. 1: *Early Years and Experiences* (1922), Vol. 2: *Central Park* (1928), Putnam's Sons, New York, London, Knickerbocker Press.

Olsen, D. J. (1976) *The Growth of Victorian London*, London, Batsford.

Osborn, F. J. (1946) *New Towns after the War*, London, J. M. Dent.

Osborn, F. J. Whittick (1969) *The New Towns the Answer to Megalopolis*, London, Leonard Hill.

Parker, B. (1923) Earswick, Yorkshire, Lecture given before the Town Planning Institute, 16 October in York (Garden City Museum Letchworth Collection of Parker's Papers).

Parker, B. (1932) Wythenshawe Estate, Manchester, *Garden Cities and Town Planning*, Vol. 22, pp. 41–5.

Parker, B. (1932/33) Highways, Parkways and Freeways, *Town and Country Planning*, Vol. 1, pp. 38–42.

Parker, B. (1937) Site Planning as exemplified at New Earswick, *The Town Planning Review*, Vol. 17, pp. 1–23.

Parker, B. and R. Unwin (1901) *The Art of Building a Home*, London, New York, Bombay, Longman, Green & Co.

Parker, B. and R. Unwin (1902) *Cottage Plans and Common Sense*, London, the Fabian Society, Buxton.

Parker, B. and R. Unwin (1903) *Cottages near a Town*, Catalogue of the Northern Art Workers' Guild Exhibition, Manchester.

Parker, B. (possibly 1940) Not officially published.

Peters, P. (1977) *Fußgängerstadt, Fußgerechte Stadtplanung und Stadtgestaltung*, München, Callwey Verlag.

Petsch, J. (1976) *Baukunst und Stadtplanung im Dritten Reich*, München, Wien, Hanser.

Pfankuch, P. and M. Schneider (1977) Von der funktionellen zur futorischen Stadt, Planen und Bauen in Europe von 1913–1933, in D. Reimer (ed.), *Tendenzen des Zwanziger Jahre*, 15. Europäische Kunstausstellung, Akademie der Künste, Berlin, Dietrich Reimer, pp. 2/1–2/48.

Pfundt, K. (1969) *Unfälle mit Fußgänger*, Köln, HUK-Verband.

Pfundt, K., V. Meewes and K. Eckstein (1975) *Verkehrssicherheit neuer Wohngebiete – Unfall-und Strukturanalyse von 10 Neubaugebieten*, Köln, HUK-Verband.

Pirath, C. (1948) *Die Verkehrsplanung: Grundlagen und Gegenwartsprobleme*, Stuttgart, Julius Hoffmann.

Plowden, S. (1987) *A Case For Traffic Restraint in London*, London, The London Centre for Transport Planning.

Plowden, W. (1971) *The Motor Car and Politics 1896–1970*, London, Sydney, Toronto, Bodley Head.

Pressman, N. (1981, ed.) International Experiences in Creating Livable Cities, *Journal of Urban and Environmental Affairs*, Vol. 13.

Prestwood-Smith, P. (1979) Area Control, *PTRC, Traffic and Environmental Management*, London, pp. 37–49.

Purdom, C. P. (1925) *The Building of Satellite Towns*, London, Dent.

Rasp, H.-P. (1981) *Eine Stadt für tausend Jahre – München – Bauten und Projekte für die Hauptstadt der Bewegung*, München, Süddeutscher Verlag.

Reichow, H. B. (1941) Grundsätzliches zum Städtebau im Altreich und im neuen deutschen Osten, in A. Teut (1967, ed.), *Architektur im Dritten Reich 1933–1945*, Berlin, Frankfurt/M., Wien, Ullstein, pp. 332–41.

Reichow, H. B. (1959) *Die autogerechte Stadt*, Ravensburg, Otto Maier.

Reimer, D. (1977) Tendenzen der Zwanziger Jahre: *see* Pfankuch and Schneider (1977).

Reindl, J. (1961) *Die Stadt ohne Verkehrsprobleme*, München.

Reps, J. W. (1965) *The Making of Urban America, A History of City Planning in the United States*, Princeton, New Jersey, Princeton University Press.

Reynolds, J. F. J. (1904) *General Notes on the London Traffic Problem*, Civil and Mechanic Engineers Society, London.

Rimpl, H. (1939) Die Stadt der Hermann Göring Werke, in A. Teut (1967, ed.), *Architektur im Dritten Reich 1933–1945*, Berlin, Frankfurt/M., Wien, Ullstein.

Ritter, P. (1964) *Planning for Man and Motor*, Oxford, London, New York, Pergamon Press.

Roberts, J. (1981) *Pedestrian Precincts in Britain*, London, TEST.

Robinson, C. M. (1916) *City Planning with Special Reference to the Planning of Streets and Lots*, New York, London, Putnam's Sons.

Royal Commission on London Report (1905) *Houses of Parliament, Reports from Commissioners, Inspectors and Others*, Vol. 30, London.

Sachs, W. (1984) *Die Liebe zum Automobil, Ein Rückblick in die Geschichte unserer Wünsche*, Hamburg, Rowohlt.

Schacht, von (1938) Der Radwegebau, in B. Rentsch (ed.), *Elsners Taschenbuch für den Straßenbau*, Berlin, Otto Elner.

Schaechterle, K. H. (1970) *Verkehrsentwicklung in deutschen Städten*, München, ADAC.

Schäffer, F. (1970) *The New Town Story*, London, MacGibbon & Kee, 2nd edn 1982.

Schaffer, D. (1982) *Garden Cities for America*, Philadelphia, Temple University Press.

Scheibe, W. (1925) Der Verkehrsraum des Stadtkerns, *Die Baugilde*, Vol. 7, pp. 1318–22.

Schlumms, J. (1955) Neuzeitliche Straßenverkehrstechnik in Deutschland, *Verkehr und Technik*, Vol. 8, pp. 67–9.

Schlumms, J. (1961) Wie ist der städtische Straßenverkehr zu bewältigen?, in Kröner Verlag (ed.), *Straßenverkehr – Problem ohne Ausweg?* Stuttgart, pp. 137–52.

Schneider, C. (1979) *Stadtgründung im Dritten Reich Wolfsburg und Salzgitter*, München, Heinz Moos Verlag.

Schubert, H. (1966) Berücksichtigung des Fußgängers in der Verkehrsplanung, *Der Städtebund*, Vol. 21, pp. 25–9.

Schubert, H. (1967) Planungsmaßnahmen für den Fußgängerverkehr in den Städten, *Städtebau und Verkehrstechnik*, Heft 56.

Scott, M. (1969) *American City Planning since 1890*, Berkeley, Los Angeles, University of California Press.

Scully, V. (1969) *American Architecture and Urbanism*, London, Thames & Hudson.

Sharp, T. (1932) *Town and Countryside: Some Aspects of Urban and Rural Development*, London, Oxford University Press.

Sharp, T. (1940) *Town Planning*, Harmondsworth, Penguin.

Sharp, T. (1945) *Cathedral City, A Plan for Durham*, London, Architectural Press.

Sharp, T. (1946) *Exeter Phoenix, A Plan for Rebuilding*, London, Architectural Press.

Sharp, T. (1948) *Oxford Replanned*, London, Architectural Press.

Sharp, T. (1949a) *Newer Sarum, A Plan for Salisbury*, London, Architectural Press.

Sharp, T. (1949b) *Georgian City, A Plan for the Preservation and Improvement of Chichester*, Brighton, Southern Publishing Co Ltd.

Sitte, C. (1965) *City Planning According to Artistic Principles*, London, Phaidon.

Smeed, A. J. (1961) *The Traffic Problem in Towns*, Manchester Statistical Society.

Smeed, R. J. (1963) The Road Space Required for Traffic in Towns, *The Town Planning Review*, Vol. 33, pp. 276-92.

Smeed, R. J. (1963) Discussion: Traffic in Towns, in British Road Federation (ed.), *People and Cities Report of the 1963 London Conference*, pp. 27-9.

Speer, A. (1970) *Inside the Third Reich*, London, Weidenfeld & Nicolson.

Stadtplanungsamt Essen (1971) *Fußgängerstraßen in Essen*, Essen.

Starkie, D. (1982) *The Motorway Age Road and Traffic Policies in Post-War Britain*, Oxford, New York, Pergamon Press.

Stein, C. S. (1925) Dinosaur Cities, in C. Sussman (1976, ed.), *Planning the Fourth Migration*, Cambridge, Mass., London, MIT Press, pp. 65-74.

Stein, C. S. (1951) *Towards New Towns for America*, Liverpool, Liverpool University Press (2nd edn, 1958), first published in Britain in *The Town Planning Review*, 1949 and 1950, Vol. 20, 21, pp. 205-99.

Stein, C. S. (1958) - see Stein, 1951.

Stevenson, J. T. (1979) Traffic in the Housing Environment, *PTRC*, Traffic and Environmental Management, pp. 253-65.

Stübben, H. J. (1890) *Der Städtebau*, Darmstadt, Bergsträsser.

Stübben, H. J. (1902) *Die Bedeutung der Bauordnungen und Bebauungspläne für das Wohnungswesen*, Göttingen, Vanderhoeck and Ruprecht.

Stübben, H. J. (1906) Planning of Streets, *The Journal of RIBA*, Vol. 13, pp. iv-v.

Stübben, H. J. (1910) Neue Fortschritte im Deutschen Städtebau, *The Royal Institute of British Architects, Transactions of the Town Planning Conference*, Oct. 1910, pp. 306-12.

Stübben, H. J. (1925) Sonderbericht von der internationalen New Yorker Städtebautagung April-Mai 1925, *DBZ, Stadt und Siedlung*, Vol. 59, pp. 186-9.

Stübben, H. J. (1929) Kraftwagenverkehr und Stadtgestaltung, *DBZ, Stadt und Siedlung*, Vol. 63, pp. 85-8.

Sussman, C. (1976) Introduction, in C. Sussman (ed.), *Planning the Fourth Migration*, Cambridge, Mass., London, MIT Press, pp. 1-45.

Sutcliffe, A. (1981) *Towards the Planned City*, Oxford, Basil Blackwell.

Sutton, S. B. (1971, ed.), *Civilizing American Cities: A Selection of Frederick Law Olmsted's Writings on City Landscape*, Cambridge, Mass., London, MIT Press.

Tamms, F. (1961) *Über die Untersuchungen für einen Generalverkehrsplan der Stadt Düsseldorf, Bericht vor dem Rat der Stadt Düsseldorf*, Düsseldorf.

Tamms, F. (1963) Townscaping Düsseldorf, in British Road Federation (ed.), *People and Cities, Report of the 1963 London Conference*, December, pp. 114-31.

Taut, B. (1914) Die Gartenstadt Falkenberg b. Berlin, Der neue Bebauungsplan, *Gartenstadt*, Vol. 8, pp. 49-51.

Taut, B. (1919) *Die Stadtkrone*, Jena, Eugen Diedrichs.

TEST (1984) *The Company Car Factor*, London, TEST.

TEST (1985) *The Accessible City*, London, TEST.

TEST (1988) *Quality Streets*, London, TEST.

Tetlow, J. and A. Goss (1965) *Homes, Towns and Traffic*, London, Faber & Faber.

Teut, A. (1967) *Architektur im Dritten Reich 1933–1945*, Berlin, Frankfurt/M. Wien, Uhlstein.

Topp, H. (1973) Untersuchungen über Begehungshäufigkeiten, Ph. D. Institut für Verkehrsplanung und Verkehrstechnik der Technischen Universität Darmstadt, Darmstadt.

Tripp, A. (1950) *Road Traffic and Its Control*, London, Edward Arnold & Co, 1st edn, 1938.

Tripp, A. (1951) *Town Planning and Road Traffic*, London, Edward Arnold & Co, 1st edn, 1942.

TRRL (1988) *Urban Safety Project, 2. Interim Results for Area-wide Schemes*, Research Report 154, Crowthorne.

Uhlig, K. (1979) *Die fußgängerfreundliche Stadt*, Stuttgart, Gerd Hatje.

Unwin, E. (1931) The International Housing and Town Planning Congress: Berlin 1931, *The Journal of the RIBA*, Vol. 38, pp. 650–1.

Unwin, R. (1904) The Improvement of Towns, Paper read at the *Conference of the National Union of Women Workers of Great Britain and Ireland*, 8 November.

Unwin, R. (1906) The Planning of the Residential Districts in Town, *The Journal of the RIBA*, Vol. 13, pp. vi–vii.

Unwin, R. (1909; 1971) *Town Planning in Practice*, Benjamin Blom, Inc., New York.

Unwin, R. (1910) The City Development Plan, in The Royal Institute of British Architects (ed.), *Transaction of the Town Planning Conference*, pp. 247–65.

Unwin, R. (1911) Karlsruhe, *The Architectural Review*, Vol. 29, pp. 54–8.

Unwin, R. (1912) *Nothing Gained by Overcrowding*, London.

Unwin, R. (1914) Roads and Streets, *The Town Planning Review*, Vol. 5, pp. 31–8.

Unwin, R. (1921) American Architecture and Town Planning, *The Journal of the RIBA*, Vol. 29, pp. 1–8.

Unwin, R. (1923) Higher Building in Relation to Town Planning, *The Journal of the RIBA*, Vol. 31, pp. 1–26.

Unwin, R. (1924) Town Planning in Relation to Land Values, *The Town Planning Review*, Vol. 5, pp. 107–14.

Van der Broek, Bakema (1956) The Lijnbaan at Rotterdam, *The Town Planning Review*, Vol. 27, pp. 21–6.

Van Doesburg, T. (1929) Die Verkehrsstadt, *Architektur der Gegenwart*, Vol. 3, pp. 4–10.

Vedral, B. (1985) *Altstadtsanierung und Wiederaufbauplanung in Freiburg i, Br. 1925–1951*, Freiburg, Schillinger Verlag.

Verkehrstechnik (1939, ed.), Kraftfahrzeugdichte in den wichtigsten Ländern, *Verkehrstechnik*, Vol. 20, p. 199.

Warnemünde, Reg. Baumeister (1929) Danzig: Städtebauliche Gedanken, *DBZ, Stadt und Siedlung*, Vol. 63, pp. 8–9.

Watson, J. P. and P. Abercrombie (1943) *A Plan for Plymouth*, Plymouth, Underhill Ltd.

Webb, S. and B. (1913) *English Local Government: The Story of the King's Highway*, London, New York, Calcutta and Bombay, Longmans, Green and Co.

Wehner, B. (1959) Stadt und Verkehr, in K. Otto (ed.), *Die Stadt von Morgen, Gegenwartsprobleme für alle*, Berlin, Gebrüder Mann, pp. 62–72.

Williams, R. (1975) *The Country and the City*, St Albans, Paladin.

Winnemöler, H. H. (1982) *Verkehrsberuhigung*, Stuttgart, Deutsche Verlagsanstalt.

Wohl, P. and A. Albitreccia (1935) *Road and Rail in 40 Countries*, London, Oxford University Press.

Wolf, P. (1928) Städtebauliche Reiseeindrücke in den Vereinigten Staaten von Amerika, *DBZ, Stadt und Siedlung*, Vol. 62, pp. 861–86.

Wood, A. A. (1967) *City of Norwich*, Draft Urban Plan, Norwich.

Wright, H. (1925) The Road to good Houses, in C. Sussman (1976, ed.), *Planning the Fourth Migration*, Cambridge, Mass., London, MIT Press, pp. 121–8.

Wright, H. (1935) *Rehousing Urban America*, New York, Columbia University Press.

Wurzer, R. (1974) Die Gestaltung der deutschen Stadt im 19. Jahrhundert, in L. Grote (ed.), *Die deutsche Stadt im 19. Jahrhundert: Stadtplanung und Baugestaltung im industriellen Zeitalter*, Studien zur Kunst des 19. Jahrhunderts, Band 24, München, Prestel Verlag, pp. 9–32.

Yago, G. (1984) *The Decline of Transit*, Cambridge, London, New York, New Rochelle, Melbourne, Sydney, Cambridge University Press.

Index